AF595044

Flood Early Warning and Risk Modelling

Flood Early Warning and Risk Modelling

Editors

Marina Iosub
Andrei Enea

MDPI • Basel • Beijing • Wuhan • Barcelona • Belgrade • Manchester • Tokyo • Cluj • Tianjin

Editors

Marina Iosub
Integrated Center of Environmental Science Studies in the North Eastern Region—CERNESIM, Sciences Department, Interdisciplinary Research Institute; Faculty of Geography and Geology
Alexandru Ioan Cuza University
Iasi
Romania

Andrei Enea
Faculty of Geography and Geology
Alexandru Ioan Cuza University of Iasi
Iasi
Romania

Editorial Office
MDPI
St. Alban-Anlage 66
4052 Basel, Switzerland

This is a reprint of articles from the Special Issue published online in the open access journal *Hydrology* (ISSN 2306-5338) (available at: www.mdpi.com/journal/hydrology/special_issues/flood_risk_modelling).

For citation purposes, cite each article independently as indicated on the article page online and as indicated below:

LastName, A.A.; LastName, B.B.; LastName, C.C. Article Title. *Journal Name* **Year**, *Volume Number*, Page Range.

ISBN 978-3-0365-3778-8 (Hbk)
ISBN 978-3-0365-3777-1 (PDF)

Contents

About the Editors

Marina Iosub

PhD Marina Iosub is a Scientific Researcher at L4-CERNESIM-ICI and Associate Assistant at Faculty of Geography and Geology, Alexandru Ioan Cuza University of Iasi. She obtained the title of Doctor of Geography, in 2017, at "Alexandru Ioan Cuza"University of Iasi, Romania. PhD Iosub Marina's research interests are related to hydrological risk analysis and flood simulations. Research directions focus on interdisciplinary research on natural hazards, GIS used in the analysis of hydrological risks by applying different methods of spatial modeling and calculating the probabilities of their occurrence, mathematical modeling of floods using specific software, statistical analysis of climate change, and the creation of risk maps. Another direction of research is given by remote sensing analyses of extreme hydrological phenomena, but also of the response of vegetation to various extreme climatic phenomena.

Andrei Enea

PhD Andrei Enea is an Assistant Professor at Faculty of Geography and Geology, Alexandru Ioan Cuza University of Iasi, and he obtained his PhD in Geography in 2017 at "Alexandru Ioan Cuza"University of Iasi, Romania. His research fields of interests are related to GIS (Geographic Informational Sistems), hydrology, remote sensing, drone imagery, and land use analysis. The research directions focus on performing vector and raster analyses at different spatial scales, flood modeling, quantifying flood risk (and other phenomena), satellite images analyses, or using aerial drone photogrammetry for modeling areas, through SFM (Structure From Motion) techniques, generating terrain models, performing analyses related to land use, land cover and their dynamics. Furthermore, another field of interest is using multi-criteria analysis, on small- and large-scale areas, for identifying spatial dependencies both on isolated, and on more complex phenomena, from a temporal resolution perspective. Lastly, PhD Andrei Enea studies spatial model creation using GIS functions, to perform natural and anthropic analysis.

Editorial

Flood Early Warning and Risk Modelling

Marina Iosub [1,2,*] and Andrei Enea [2,*]

1 Integrated Center of Environmental Science Studies in the North Eastern Region—CERNESIM, Sciences Department, Interdisciplinary Research Institute, Alexandru Ioan Cuza University of Iași, 700506 Iași, Romania
2 Department of Geography, Faculty of Geography and Geology, Alexandru Ioan Cuza University of Iași, 700505 Iași, Romania
* Correspondence: marina.iosub@uaic.ro (M.I.); andrei.enea@uaic.ro (A.E.)

1. Introduction

The evolution of mankind during the last 2 centuries has generated an ever growing thrive for increased production, for the need to create novel means to generate energy and for society to change into a more consumerism-oriented version. All of these aforementioned arguments induce repeated and frequently irreversible repercussions on nature, and through constant alterations of the environment, significant global climate changes have been observed especially in the most recent times. During the last half of the century, all of these climatic changes have induced, as direct effects, the exponential rise in the number of natural disasters, and this increase is also emphasized by the ever-growing intensification of these phenomena, with devastating effects [1]. Out of all these extreme natural phenomena, floods are statistically considered some of the most damage-inducing, with an ever-more frequent recurrence time period, and can originate from fluvial sources, from extreme torrential rainfall, from coastal floods, or even due to underground causes. On a global scale, out of the entirety of registered events that took place since 1900, floods account for 25% of the events that occurred since 1900, 38% of the events registered since 1960, and 45% for events that took place since 2000 [2]. The population affected by this type of risk has dramatically decreased since the 1960′s, but the economic damages from floods, as a shared portion of the GDP at a global level, has risen from 0.02% (during the timespan between 1960–1969), up to 0.05% (between 2010–2019) [3]. The economic implications (costs of infrastructure, insurances, and even social-related issues) induced on the local communities are expected to increase even more in the following years, due to the expansion of the urban built-up areas, economic growth, and especially climate change [4,5].

Citation: Iosub, M.; Enea, A. Flood Early Warning and Risk Modelling. *Hydrology* **2022**, *9*, 57. https://doi.org/10.3390/hydrology9040057

Received: 29 March 2022
Accepted: 29 March 2022
Published: 31 March 2022

The fact that natural phenomena, such as floods, induce tremendous amounts of damage, both from an economic and social perspective, has determined scientists to address this issue, and to search for potential solutions to identify areas, exposed to flood risk. In order to help mitigate this, multiple mathematical modelling algorithms were developed, with a variety of functions, such as calculating flow rate values, estimating possible recurring flow rates, mapping of flood risk, generating scenarios related to flood extent layers, warning and delay times for flash floods etc. [6,7]. Most of the aforementioned were mainly developed during the last 5 decades, justified by the increase in computational capabilities and in interest towards mitigating the negative impact of future events [2,3]. The main, general purpose they serve, is to describe the manner in which a drainage basin responds to sudden changes in flow rates, the increased values in torrential rainfall and estimate as precisely, as possible, the runoff, or even the groundwater related parameters, if required. Another aspect flood models address is the probability extent of floods along the main channel and floodplain for the main, collector river.

In order to mathematically model a flood event or simulation, significant temporal resources and labor have to be dedicated towards generating and validating the final results, therefore raising numerous technical challenges, associated to the complete characterization and cartographic representation of several physical and hydrodynamic parameters, such as: the relationship between rainfall and runoff, the heterogenic distribution of precipitation, the morphometric and pedological aspects in the study area etc. The more these parameters are better calibrated, the significantly better the final results related to the hydrological risk are, in accordance with the in-field situation [8,9]. Recent progress in the modern techniques, based on long, representative datasets are related to the increased availability of these kinds of methods, in comparison to the classic hydrodynamic analysis methodologies, which aid in the real-time forecasting of flood events. Novel detection systems (such as remote sensing techniques), in relation to modern data acquisition means, and backed by experimental techniques, provide new opportunities for calibration, validation, testing and improving the flood models [10–12].

The heterogenic manner through which the issue of flood risk is addressed, at the scientific, quantifiable level has determined us to compile a Special Issue which would promote publications that address flood analysis and apply some of the most novel inundation prediction models, as well as various hydrological risk simulations related to floods, that will enhance the current state of knowledge in the field as well as lead toward a better understanding of flood risk modeling. Furthermore, the current Special Issue will address the temporal aspect of flood propagation, including alert times, warning systems, flood time distribution cartographic material, and the numerous parameters involved in flood risk modeling.

2. Contributed Papers

The current Special Issue has managed to compile a total number of 6 published papers, with topics varying significantly, from the application of hydrological models specific to river valleys, to the study of entire drainage basins, by using 1D methods of analysis, or even 2D methods, with the help of virtual machines, and statistical data processing techniques, related to flood events and last but not least, flood early warning systems.

The article *"Towards Coupling of 1D and 2D Models for Flood Simulation—A Case Study of Nilwala River Basin, Sri Lanka"* by Dhanapala, L.; Gunarathna, M.H.J.P.; Kumari, M.K.N.; Ranagalage, M.; Sakai, K.; Meegastenna, T.J. describes the coupling method of two hydraulic models, in order to mathematically simulate floods on the valley of Nilwala River, in Sri Lanka. The methods used are HEC-HMS for the 1D analysis, while for 2D hydrological modelling, the method applied is iRIC. The model has been validated and calibrated using data from 3 selected events that have taken place during 2017–2019-time frame. The generated results have been validated to have an overall accuracy of 81.5%. Considering the novelty of the applied methodology, the authors have proposed that this method should be applied on more recorded hydrological events, in order to help emphasize the forecasting capabilities of the model [13].

One of the most recent large-scale hydrotechnical issues of current times is related to dams, and their potential risk to fail catastrophically. Considering that water resources are an ever-increasing global issue, governments have looked for different types of solutions to identify and prevent dam failure. The severity of the topic is emphasized by the fact that most current dams have been built at the beginning of the 20th century, which, correlated to the generalized state of degradation of most dams, corresponds to the continuous aging of these types of constructions. The current study *"Flood Mapping from Dam Break Due to Peak Inflow: A Coupled Rainfall–Runoff and Hydraulic Models Approach"* by Tedla, M.G.; Cho, Y.; Jun, K. addresses the dam break simulation of Kesem Dam, from Ethiopia, using the coupling 1D rainfall-runoff method HEC-HMS, which was used to predict flood volume, peak rates, and the runoff hydrograph, as well as the 2D hydraulic model HEC-RAS, used to simulate a piping dam break. Final results indicate the predisposition of the dam to flood risk through partial or complete breaking, for flow rates corresponding to recurrence periods between 100–200 years. The resulting maps from the 2D simulations aid in the identification of risk exposed areas downstream of dams [14].

For this particular study: *"Hydrological and Hydraulic Flood Hazard Modeling in Poorly Gauged Catchments: An Analysis in Northern Italy"* by Aureli, F.; Mignosa, P.; Prost, F.; Dazzi, S., a method for hydraulic 2D simulation has been applied, through the PARFLOOD algorithm, with the purpose of generating a flood hazard map in the poorly gauged drainage basins, with the study area located along the Chiavenna river valley in Italy. The article is divided into two methodological sections: in the first part, there is an analysis of the regional flood frequency from where the synthetic design hydrographs are derived, which are also imposed as upstream boundary conditions for 2D modeling. In the second part, 2D flood simulations are performed for 2 scenarios: a real one, in which 20 bridges are taken into account as cross sections; and a hypothetical one, in which bridges are not taken into account. The 20 sections have a role in indicating whether or not they influence the flood manifestation, depending on the way in which the bridges are dimensioned and located across the riverbed. Therefore, it was concluded that many of them influence the flood depth and the extent of flooded areas upstream of the bridges. The final results are the more important, the easier they are to be included in the design of road infrastructure and hydrotechnical builds for civil protection [15].

Understanding how floods manifest introduced the need to create models that aid in the efficient taking of decisions during the early stages of flood occurrences, or even before they manifest in the inhabited areas. The current study *"Flood Early Warning Systems Using Machine Learning Techniques: The Case of the Tomebamba Catchment at the Southern Andes of Ecuador"* by Muñoz, P.; Orellana-Alvear, J.; Bendix, J.; Feyen, J.; Célleri, R. addresses the comparative processes and results for 5 machine learning algorithms designed to simulate the flood water quantity that is associated to runoff due to short term floods, inside mountainous drainage basins of small or medium sizes. The models have been tested to emphasize the results in a clear, easily understandable form, by any non-expert person, in 3 categories (No-Alert, Pre-Alert and Alert). One of the suggested limitations of the applied methodology is that it offers very good performance only for the up to the 6-hour limit for lead times for forecasting result, as generated by the algorithms [16].

In the context of ever more frequent catastrophes and exposure of society to their impact, the need to differentiate between forecasting and nowcasting gathers more and more attention. Furthermore, the need to quickly and efficiently disseminate information regarding the manifestation of floods has determined scientists to develop and apply increasingly more complex simulation models. The HAND model has been applied in representative areas in Iowa in the article *"Real-Time Flood Mapping on Client-Side Web Systems Using HAND Model"* by Hu, A. and Demir, I., and have managed to prove the high dependability and efficiency of this model, especially when paired with web-platforms. Local authorities can use this approach to identify the areas exposed to flood risk, without requiring trained personnel to interpret scientific data, during torrential events, resulting in floods. Finally, one of the main advantages of this method is the minimal usage of datasets originating from external sources [17].

Catastrophic events have always had a two-sided approach, between researchers, on one side, and the target population, on the other side, therefore implying the need to quickly disseminate useful information via an easily accessible platform. ESRI StoryMaps provides such a possibility, as proved by the authors Oubennaceur, K.; Chokmani, K.; El Alem, A.; Gauthier, Y. of the study *"Flood Risk Communication Using ArcGIS StoryMaps"*. For this particular study, a StoryMap has been generated in the drainage basin of Petite-Nation River, which suffers from flood damage on a frequent basis. Both vector and raster data has been integrated into the StoryMap, to more accurately represent the on-site situation, on a high level of detail. As a case study, the manuscript emphasizes the current flood-related situation, regarding risk management, but also includes predictive data which is related to future climate change [18].

3. Conclusions

The current Special Issue regarding Flood Early Warning and Risk Modelling, has successfully addressed different approaches of studying the way that flash floods manifest in the differently-sized drainage basins. Furthermore, these diverse papers have used 1D and 2D modelling methods, as well as comparative analysis between several machine learning algorithms, or hydraulic models, such as HAND model, which uses more morphometric based parameters. The emphasized has been also put on disseminating information on web-based virtual platforms, such as StoryMaps.

Author Contributions: The two Guest Editors (M.I. and A.E.) equally shared the work of soliciting, editing, and handling the contributions made to this Special Issue. All authors have read and agreed to the published version of the manuscript.

Funding: The creation of this Special Issue did not receive external funding.

Institutional Review Board Statement: Not applicable.

Informed Consent Statement: Informed consent was obtained from all subjects involved in the study.

Data Availability Statement: Not applicable.

Acknowledgments: We would like to acknowledge the efforts of all authors that contributed to the Special Issue.

Conflicts of Interest: The authors declare no conflict of interest.

References

1. European Commission. Directorate-General for Environment, Impact of Climate Change on Floods: Survey Findings and Possible Next Steps to Close the Knowledge and Implementation Gap: Final a Survey Based Study, Publications Office, 2021. Available online: https://data.europa.eu/doi/10.2779/932572 (accessed on 15 March 2022).
2. EM-DAT. *The Emergency Events Database*; Université Catholique de Louvain (UCL)-CRED: Brussels, Belgium, 2016. Available online: www.emdat.be (accessed on 15 March 2022).
3. Our World in Data. *Natural Disaster*; Global Change Data Lab: Wales, UK, 2014. Available online: https://ourworldindata.org (accessed on 20 March 2022).
4. Liu, H.; Wang, Y.; Zhang, C.; Chen, A.; Fu, G. Assessing real options in urban surface water flood risk management under climate change. *Nat. Hazards* **2018**, *94*, 1–18. [CrossRef]
5. Ridolfi, E.; Mondino, E.; Di Baldassarre, G. Hydrological risk: Modeling flood memory and human proximity to rivers. *Hydrol. Res.* **2021**, *52*, 241–252. [CrossRef]
6. Enea, A.; Urzica, A.; Breaban, I.G. Remote sensing, GIS and HEC-RAS techniques, applied for flood extent validation, based on Landsat imagery, LiDAR and hydrological data. Case study: Baseu River, Romania. *J. Environ. Prot. Ecol.* **2018**, *19*, 1091–1101.
7. Iosub, M.; Minea, I.; Chelariu, O.E.; Ursu, A. Assessment of flash flood susceptibility potential in Moldavian Plain (Romania). *J. Flood Risk Manag.* **2020**, *13*, e12588. [CrossRef]
8. Jain, S.K.; Mani, P.; Jain, S.K.; Prakash, P.; Singh, V.P.; Tullos, D.; Kumar, S.; Agarwal, S.P.; Dimri, A.P. A Brief review of flood forecasting techniques and their applications. *Int. J. River Basin Manag.* **2018**, *16*, 329–344. [CrossRef]
9. Huţanu, E.; Mihu-Pintilie, A.; Urzica, A.; Paveluc, L.E.; Stoleriu, C.C.; Grozavu, A. Using 1D HEC-RAS Modeling and LiDAR Data to Improve Flood Hazard Maps Accuracy: A Case Study from Jijia Floodplain (NE Romania). *Water* **2020**, *12*, 1624. [CrossRef]
10. Mei, Y.; Anagnostou, E. A hydrograph separation method on information from rainfall and runoff records. *J. Hydrol.* **2015**, *523*, 639–649. [CrossRef]
11. Zhou, S.; Wang, Y.; Li, Z.; Chang, J.; Guo, A. Quantifying the Uncertainty Interaction between the Model Input and Structure on Hydrological Processes. *Water Resour. Manag.* **2021**, *35*, 3915–3935. [CrossRef]
12. Falter, D.; Dung, N.V.; Vorogushyn, S.; Schröter, K.; Hundecha, Y.; Kreibich, H.; Apel, H.; Theisselmann, F.; Merz, B. Continuous, large-scale simulation model for flood risk assessments: Proof-of-concept. *J. Flood Risk Manag.* **2016**, *9*, 3–21. [CrossRef]
13. Dhanapala, L.; Gunarathna, M.H.J.P.; Kumari, M.K.N.; Ranagalage, M.; Sakai, K.; Meegastenna, T.J. Towards Coupling of 1D and 2D Models for Flood Simulation—A Case Study of Nilwala River Basin, Sri Lanka. *Hydrology* **2022**, *9*, 17. [CrossRef]
14. Tedla, M.G.; Cho, Y.; Jun, K. Flood Mapping from Dam Break Due to Peak Inflow: A Coupled Rainfall-Runoff and Hydraulic Models Approach. *Hydrology* **2021**, *8*, 89. [CrossRef]
15. Aureli, F.; Mignosa, P.; Prost, F.; Dazzi, S. Hydrological and Hydraulic Flood Hazard Modeling in Poorly Gauged Catchments: An Analysis in Northern Italy. *Hydrology* **2021**, *8*, 149. [CrossRef]

16. Muñoz, P.; Orellana-Alvear, J.; Bendix, J.; Feyen, J.; Célleri, R. Flood Early Warning Systems Using Machine Learning Techniques: The Case of the Tomebamba Catchment at the Southern Andes of Ecuador. *Hydrology* **2021**, *8*, 183. [CrossRef]
17. Hu, A.; Demir, I. Real-Time Flood Mapping on Client-Side Web Systems Using HAND Model. *Hydrology* **2021**, *8*, 65. [CrossRef]
18. Oubennaceur, K.; Chokmani, K.; El Alem, A.; Gauthier, Y. Flood Risk Communication Using ArcGIS StoryMaps. *Hydrology* **2021**, *8*, 152. [CrossRef]

MDPI

Article

Towards Coupling of 1D and 2D Models for Flood Simulation—A Case Study of *Nilwala* River Basin, Sri Lanka

Lanthika Dhanapala [1], M. H. J. P. Gunarathna [1,*], M. K. N. Kumari [1], Manjula Ranagalage [2], Kazuhito Sakai [3] and T. J. Meegastenna [4]

1 Department of Agricultural Engineering and Soil Science, Faculty of Agriculture, Rajarata University of Sri Lanka, Anuradhapura 50000, Sri Lanka; lanthika_d@agri.rjt.ac.lk (L.D.); nadeeka@agri.rjt.ac.lk (M.K.N.K.)
2 Department of Environmental Management, Faculty of Social Sciences and Humanities, Rajarata University of Sri Lanka, Mihintale 50300, Sri Lanka; manjularanagalage@ssh.rjt.ac.lk
3 Faculty of Agriculture, University of the Ryukyus, 1 Senbaru, Nishihara-cho, Okinawa 903-0213, Japan; ksakai@agr.u-ryukyu.ac.jp
4 Irrigation Department, Colombo 00700, Sri Lanka; adgirdc@irrigation.gov.lk
* Correspondence: janaka78@agri.rjt.ac.lk

Citation: Dhanapala, L.; Gunarathna, M.H.J.P.; Kumari, M.K.N.; Ranagalage, M.; Sakai, K.; Meegastenna, T.J. Towards Coupling of 1D and 2D Models for Flood Simulation—A Case Study of *Nilwala* River Basin, Sri Lanka. *Hydrology* **2022**, *9*, 17. https://doi.org/10.3390/hydrology9020017

Academic Editors: Marina Iosub and Andrei Enea

Received: 10 December 2021
Accepted: 21 January 2022
Published: 25 January 2022

Publisher's Note: MDPI stays neutral with regard to jurisdictional claims in published maps and institutional affiliations.

Abstract: The *Nilwala* river basin is prone to frequent flooding during the southwest monsoon and second intermonsoon periods. Several studies have recommended coupling 1D and 2D models for flood modelling as they provide sufficient descriptive information of floodplains with greater computational efficiency. This study aims to couple a 1D hydrological model (HEC-HMS) with a 2D hydraulic model (iRIC) to simulate flooding in the *Nilwala* river basin. Hourly rainfall and streamflow data of three flood events were used for calibration and validation of HEC-HMS. The model performed exceptionally well considering the Nash–Sutcliffe coefficient, percent bias, and root mean square error. The flood event of May 2017 was simulated on iRIC using the streamflow hydrographs modelled by HEC-HMS. An overall accuracy of 81.5% was attained when the simulated extent was compared with the surveyed flood extent. The accuracy of the simulated flood depth was assessed using the observed water level at *Tudawa* gauging station, which yielded an NSE of 0.94, PBIAS of −4.28, RMSE of 0.18 and R^2 of 0.95. Thus, the coupled model provided an accurate estimate of the flood extent and depth and can be further developed for hydrological flood forecasting on a regional scale.

Keywords: *Nilwala* river basin; coupled flood modelling; HEC-HMS; iRIC

1. Introduction

A significant consequence of climate change is the increased frequency and intensity of extreme weather events [1]. Germanwatch described Sri Lanka as being among the top ten most affected countries due to climate change in 2018, with a global climate risk index of 19.0 [2]. Regional studies conducted on rainfall trends in Sri Lanka have indicated both increasing and decreasing patterns [3–5]. It was found that there was a higher tendency for short-duration high-intensity rainfall events to occur within the island [4], which often resulted in water-related natural hazards. Floods are destructive and occur more frequently than other natural disasters [6], with Asia being one of the most affected regions [7]. The Fifth Assessment Report of the IPCC states with medium confidence that the magnitude and frequency of recent floods are comparable to or surpass historical floods [1].

Sri Lanka was recently devastated by floods and landslides due to heavy southwest monsoonal rainfall in May 2017. This natural disaster led to 219 deaths, affecting approximately 230,000 families [8]. The most adversely affected districts by this disaster were Kalutara, Galle, Matara, Hambantota, and Ratnapura. Flooding occurred in towns, villages, and agricultural land bordering the *Kalu*, *Nilwala*, and *Gin* rivers. Matara city is located in the floodplain of the *Nilwala* river basin, an area prone to flooding at the onset of the southwest monsoonal rains. Matara is the administrative capital of the Matara district and is considered a commercial hub. The risk of flooding is significantly greater in urban areas, such as Matara city where impervious surfaces prevent infiltration and cause runoff to generate quickly [9].

A past study identified precipitation within the basin increases from the south to the north with an increasing elevation [10]. Thus, a significant proportion of the rainfall that contributes to flooding in the lower coastal plains occurs in the basin's upper region. There is a considerable period of time for extreme precipitation in upper *Nilwala* to accumulate into floods in the lower coastal floodplains. With the recent advances in information and communication technology, this study aims to assess the feasibility of applying a 1D–2D coupled model to simulate flooding in real-time that could provide sufficient lead time for evacuation and disaster preparation.

The Post Disaster Needs Assessment (PDNA) was conducted following the disaster in 2017 and identified several vital interventions: strengthening of the country's early warning system, the conduction of hazard risks, and vulnerability assessments were points focused on. Moreover, studies have shown that mortalities caused by floods depend on the scale or severity of the flood, the possibility of an early warning mechanism, and evacuation [11]. Computational models can be used to understand the nature of the flood and simulate hydrological responses to various control measures [12]. They can also generate flood risk maps for planning and developing the resiliency of high-risk regions [13]. The existing early warning system of the *Nilwala* river basin consists of a network of water gauging stations with extreme water levels categorized as alert, minor flood, and major flood. Flood warnings are issued based on this water level and its trend. There remains room for improvement of the system through flood modeling.

Hydraulic models can be classified dimensionally as 1D, 2D, and 3D models. Three-dimensional models were developed relatively recently, requiring more descriptive parameters that remain unavailable for most basins [14]. The simplest representation of flow in a stream is as a one-dimensional flow in a longitudinal direction. Although these models are computationally efficient, they demonstrate certain limitations; namely the topological discretization of topography as cross-sections of the floodplain [15], the subjectivity of the location and orientation of cross-sections, and the incapability of simulating lateral diffusion of the flood wave [16]. A 2D model solves for water flow equations, both in longitudinal and lateral planes by using either finite element or finite volume analysis methods [17]. They can provide additional detailed information on the floodplain and allow the visualization of flood extent, but are computationally more demanding, requiring greater processing power and storage capacity.

In this study, we coupled a 1D hydrological model to a 2D hydraulic model. The upper portion of the basin is modelled one-dimensionally to carry the flood wave to the two-dimensionally modelled coastal floodplains, which is an urban area with a high risk of flooding. The coupled model will generate comprehensive information of the floodplain with greater computational efficiency [13]. The coupled model allows one to alter the roughness coefficient of the simulation extent, provide a more detailed description of the topography, simulate overland flow, and can be easily visualized as two-dimensional maps [12]. Teng et al., [18] describes several methods to couple a 1D model with a 2D model. These include; modelling a section of the stream in 1D and the rest in 2D, coupling a 1D drainage model with a 2D flood model [19], and coupling a 1D river model with a 2D floodplain model [14,20,21]. This study uses the second approach, where Hydrologic Engineering Center—Hydrologic Modelling System (HEC-HMS) is used to simulate drainage in upper *Nilwala* and is externally coupled with the 2D hydraulic model, the Nays2DFlood solver, in the International River Interface Cooperative (iRIC).

This approach has been successfully used for flood forecasting and urban flood management. A study conducted in the Brahmani and Baitarani river delta in India by coupling SWAT and SWMM with iRIC yielded satisfactory results and was able to identify the depth and extent of inundation [19]. A similar study using a coupled model conducted in China was similarly successful and capable of assessing the effects of water conservancy structures within the simulation extent [22].

HEC-HMS is a rainfall-runoff hydrological model developed by the United States Army Corps of Engineers used extensively to simulate the hydrological response of catchments [23–25]. This model is used to estimate the runoff generated upstream of the 2D flood modeling extent and to provide the boundary conditions required for flood simulation. HEC-HMS was selected in this study due to its versatility in modeling hydrological processes and ability to model storm events. Nays2DFlood solver in iRIC is a 2D hydraulic model used for flood inundation mapping [19,26–29]. The iRIC is a recent hydraulic model that can model large flood events [30]. Owing to the nature of its recency, it has not been applied widely, but was able to simulate flooding in ungauged catchments where data are scarce. It is a free open-source software developed by Yasuyuki Shimizu and Toshiki Iwasaki of Hokkaido University, Japan.

2. Materials and Methods

2.1. Study Area

Figure 1 depicts the *Nilwala* river basin located along the southern coast of Sri Lanka. The river is 72 km in length and originates at *Panilkanda*, north-east of *Deniyaya*, at 1050 m above mean sea level [10]. The total drainage area of the river basin is 1025 km^2 [31]. Steep longitudinal slopes characterize the basin's upper region, while the lower basin consists of gentle slopes [31,32]. It passes several small towns such as *Deniyaya*, *Morawaka*, and *Akuressa* and flows into the ocean *Thotamuna* close to Matara city. The catchment is located in the wet zone, with most of the upper catchment covered by rainforests. A significant portion of the basin is used for cultivating agricultural crops such as paddy, coconut, tea, rubber, cinnamon, and citronella [32,33].

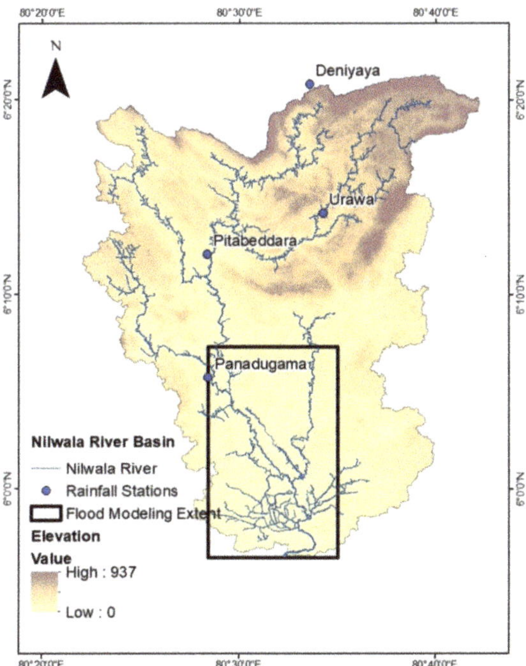

Figure 1. *Nilwala* river basin.

2.2. Data

Hourly rainfall and streamflow data required for simulation were obtained from the Irrigation Department of Sri Lanka. The *Nilwala* river has three streamflow gauging stations; *Urawa, Pitabeddara,* and *Panadugama,* from which hourly streamflow data were collected. Continuous data of the 2017 flood event were not available for *Urawa* and *Panadugama* stations. Thus, calibration and validation of HEC-HMS was undertaken solely using data obtained from *Pitabeddara*. Hourly precipitation data from four stations in upper *Nilwala; Deniyaya, Panadugama, Pitabeddara,* and *Urawa* were additionally used. The data used in this study are summarized in Table 1.

Table 1. Summary of data used in study.

Data Type	Source	Frequency	Period/Scale
Precipitation	Irrigation Department	Hourly	2017–2019
Streamflow	Irrigation Department	Hourly	2017–2019
Water level	Irrigation Department	Hourly	May–June 2017
Digital Elevation Model (DEM)	United States Geological Survey (USGS)	-	30 m

2.3. Rainfall-Runoff Modelling Using HEC-HMS

The Hydrologic Engineering Center—Hydrologic Modelling System (HEC-HMS) model was designed by the US Army Corps of Engineers as a rainfall-runoff model to determine the hydrological response of watersheds [34]. It was developed for dendritic watersheds and has a wide array of applications, including urban drainage, flood-loss reduction, flood early warning system planning, reservoir spillway design, stream restoration, flow forecasting, surface erosion, sediment routing, and systems operation [35].

The *Nilwala* basin was delineated and divided into 20 subbasins using the HEC-GeoHMS extension in ArcGIS. Terrain preprocessing and basin processing tools were used to generate the basin model file containing the drainage network and delineated basin. Next, precipitation was estimated using the Thiessen polygon method [36], and suitable weights for each subbasin were defined to create a meteorological model. The basin and meteorological models were imported into HEC-HMS. Hourly precipitation data from the *Deniyaya, Panadugama, Pitabeddara,* and *Urawa* gauging stations, and observed discharge data for the selected events were imported into the HEC-HMS interface.

HEC-HMS allows for the separate modelling of hydrological processes; loss, transformation, baseflow, and routing with several models for each process. The selection of the model for each process should be based on the catchment characteristics, data availability, and whether the simulation is event-based or continuous [25].

The Snyder unit hydrograph method [37] was selected to simulate the transform in conjunction with the deficit and constant loss method, and the recession baseflow model [38]. The lag time, t_p to be used in the Snyder unit hydrograph was defined as the time period between the centroid of excess rainfall and the peak discharge. It was calculated for each subbasin using the following relationship;

$$t_p = CC_t(LL_c)^{0.3}, \tag{1}$$

where C_t = basin coefficient, L = length of the mainstream from the outlet to the basin divide, L_c = length along the mainstream from the outlet nearest the watershed centroid and C = a conversion factor (0.75 for SI units).

The Muskingum–Cunge method [39] was used to model routing in all reaches with a Manning's n coefficient of 0.040 [40,41]. Parameters such as reach length and slope were estimated using topographical maps. A deterministic optimization approach was followed with NSE set as the objective function. Optimization trials were carried out for the deficit and constant loss model, and the recession baseflow model to determine proper parameters. Table 2 lists the parameters optimized in HEC-HMS.

Table 2. Parameters optimized in HEC-HMS model.

Parameter	Units	Optimized Value	Model
Initial deficit	mm	3.62	Deficit and constant loss
Maximum storage	mm	67.575	Deficit and constant loss
Constant rate	mm/h	6.135	Deficit and constant loss
Impervious	%	30	Deficit and constant loss
Initial discharge	$m^3/s/km^2$	0.1	Recession baseflow
Recession constant	dimensionless	0.45	Recession baseflow
Ratio to peak	dimensionless	0.25	Recession baseflow

The model was initially applied to the basin's upper region reaching up to *Pitabeddara* due to the lack of reliable streamflow data below this point. Once the model was calibrated and validated, it was then extended to the entire basin. For the simulation, three flood events were selected: May 2017, May 2018, and September 2019. The May 2018 and September 2019 events were used for calibration, while the May 2017 event was used to validate the model. The two selected events for calibration had both reported flooding, although not at the magnitude of the May 2017 flooding event.

2.4. Hydrodynamic Modelling Using Nays2DFlood Solver in iRIC

The International River Interface Cooperative (iRIC) is a free numerical simulation software consisting of various computational solvers to decipher hydrological problems. Nays2DFlood is one such solver in iRIC, which is used for flood simulation. The iRIC software platform was initiated by Professor Yasuyuki Shimizu (Hokkaido University) and Dr. Jonathan Nelson (USGS). Nays2DFlood relies on unsteady 2-dimensional plane flow simulations using boundary-fitted coordinates as the general curvilinear coordinates [42]. This solver can be applied to river basins with sparse data, especially in developing regions as it only requires the DEM without specific topographical dimensions of the river.

The equation of continuity (Equation (2)) and equations of motion (Equations (3) and (4)) of two-dimensional unsteady flows in a rectangular coordinate system are expressed as;

$$\frac{\partial h}{\partial t} + \frac{\partial (hu)}{\partial x} + \frac{\partial (hv)}{dy} = q + r, \tag{2}$$

$$\frac{\partial (hu)}{\partial t} + \frac{\partial (hu^2)}{\partial x} + \frac{d(huv)}{\partial y} = -hg\frac{\partial H}{\partial x} - \frac{\tau_x}{\rho} + D^x, \tag{3}$$

$$\frac{\partial (hv)}{\partial t} + \frac{\partial (huv)}{\partial x} + \frac{d(hv^2)}{\partial y} = -hg\frac{\partial H}{\partial x} - \frac{\tau_y}{\rho} + D^y, \tag{4}$$

where,

$$\frac{\tau_x}{\rho} = C_f u\sqrt{u^2 + v^2}, \tag{5}$$

$$\frac{\tau_y}{\rho} = C_f v\sqrt{u^2 + v^2}, \tag{6}$$

$$D^x = \frac{\partial}{\partial x}\left[v_t\frac{\partial (hu)}{\partial x}\right] + \frac{\partial}{\partial y}\left[v_t\frac{\partial (hu)}{\partial y}\right], \tag{7}$$

$$D^y = \frac{\partial}{\partial x}\left[v_t\frac{\partial (hv)}{\partial x}\right] + \frac{\partial}{\partial y}\left[v_t\frac{\partial (hv)}{\partial y}\right], \tag{8}$$

where, h = water depth, t = time, u = flow velocity in x direction, v = flow velocity in y direction, g = gravitational acceleration, H = water surface elevation, τ_x and τ_y are riverbed shear stress in x and y directions, respectively, C_f = riverbed friction coefficient, v_t = eddy viscosity coefficient, ρ = density of water, q = inflow through a box culvert, a sluice pipe or a pump per unit area, and r = rainfall [42]. These basic equations, which present in a rectangular coordinate system, are mapped to a generalized curvilinear coordinate system. The shape of the simulation mesh can be changed by accomplishing this depending on the boundary conditions [42].

The inputs required for iRIC are the DEM, satellite images, and streamflow hydrographs at the inlet points. The 30 m SRTM DEM from the United States Geological Survey (USGS) was used for the simulation, while the required background satellite images were obtained from Google Earth. Next, a calculation grid with 80 m grid size was created by defining a grid centerline representing the river's flow. Values for n_I, n_J, and W were defined, where n_I = number of divisions in the longitudinal direction, n_J = number of divisions in the transverse division, and W = grid width in the transverse direction. Subsequently, calculation conditions were set, including boundary conditions, initial conditions, roughness conditions, and an appropriate time step for computation. The inflow streams identified earlier were then located on the grid. The discharge hydrographs obtained from the HEC-HMS model were then imported into iRIC manually and set as boundary conditions.

Impermeable objects such as roads, banks, embankments, and other flood control structures were located on the grid as obstacles. Matara city was outlined as an area of the grid occupied by buildings, and its areal fraction occupied by buildings was defined. Most calculation conditions of iRIC were left as default with a numerical upwind scheme used as the finite differential method. The northern, eastern and western boundaries were set as inflow boundaries while the southern boundary was modelled as having free outflow. The resulting inundation map's accuracy was determined by comparison with a survey of the flood extent carried out by the Irrigation Department. The depth of flooding was validated using observed water level data at *Tudawa* gauging station.

2.5. Statistical Evaluation

Both models were statistically evaluated using Nash–Sutcliffe efficiency (NSE) [43], percent bias (PBIAS) [44], and root mean square error (RMSE).

The NSE indicates the goodness of fit of simulated data to observed data. It is calculated using the following equation, given by Nash and Sutcliffe (1970).

$$\text{NSE} = 1 - \frac{\sum_{i=1}^{n}(O_i - S_i)^2}{\sum_{i=1}^{J}(O_i - \overline{O})^2}, \tag{9}$$

where O_i = value of ith observation, S_i = value of ith simulation, $\overline{O}$ = mean of observations and n = total observations. NSE values range from $-\infty$ to 1, with an NSE of 1 indicating a perfect fit.

The PBIAS measures a simulated data's tendency to be greater than or lesser than the observed values [44]. A negative value of PBIAS indicates a model overestimation bias, while a positive value indicates a model underestimation bias. A PBIAS value of 0 is optimal.

$$\text{PBIAS} = \frac{\sum_{i=1}^{n} 100(O_i - S_i)}{\sum_{i=1}^{J} O_i}, \tag{10}$$

The RMSE was performed for quantifying the prediction error in terms of the units of the variable calculated by the model [45].

$$\text{RMSE} = \sqrt{\frac{\sum_{i=1}^{n}(O_i - S_i)^2}{n}}, \tag{11}$$

3. Results and Discussion

3.1. Calibration and Validation of HEC-HMS

The Irrigation Department has only recently begun collecting hourly streamflow and rainfall data from gauging stations at *Panadugama*, *Pitabeddara*, and *Urawa*. Thus, only three storm events were found with significant precipitation that resulted in downstream flooding. Of the three selected events, the May 2018 and September 2019 events were selected for model calibration. The flooding event of May 2017, considerably more intense among the events, was selected for validation and subsequently modelled using iRIC.

The Snyder unit hydrograph method for transformation produced excellent results during calibration and validation processes. A past study conducted within the *Nilwala* river basin confirmed that the Snyder transformation method provided the most accurate results [46]. Another study conducted in the *Attanagalu Oya* basin using HEC-HMS concluded that the Snyder unit hydrograph simulated flow with greater reliability than the Clark unit hydrograph and SCS Curve Number methods [23]. Moreover, the study recommends applying this model in conjunction with the deficit and constant loss model to simulate river flow in the country's wet zone [23].

The recession baseflow model in HEC-HMS simulates a catchment behavior when flow recedes exponentially and was designed primarily for event simulations [34]. Recession baseflow and the Muskingum–Cunge routing method were successfully used to predict flow in the *Nilwala* river basin [46]. The downstream regions of the basin display a characteristic gentle slope, modelled more accurately using the Muskingum–Cunge method [47]. Detailed river cross-sectional data could not be obtained for the study, and thus the parameters for the routing model were estimated using the basin model generated by HEC-GeoHMS and satellite imagery. Moreover, obtaining 8-point cross-sectional data could significantly improve model performance. Alternatively, a trapezoidal shape design was specified for the routing channels.

During optimization, it was understood that the loss parameters influenced the peak of discharge, while the baseflow parameters affected the initial flow and recession curve. The transformation and routing processes influenced the time of the peak. After several optimization trials, the model was adequately calibrated for selected flood events.

Figure 2a depicts the observed and simulated hydrographs for the May 2018 event, while Figure 2b shows the September 2019 flood event. The May 2018 event provided the values of 0.93 NSE and 0.3 RMSE (Table 3), categorized as being excellent according to the criteria suggested by Moriasi et al., [48] for monthly hydrological simulations. The September 2019 event produced an NSE value of 0.746 after calibration and can be categorized as useful while having an RMSE value of 0.5. The PBIAS for both events was negative, −10.95% and −4.82%, respectively, indicating an overestimation bias.

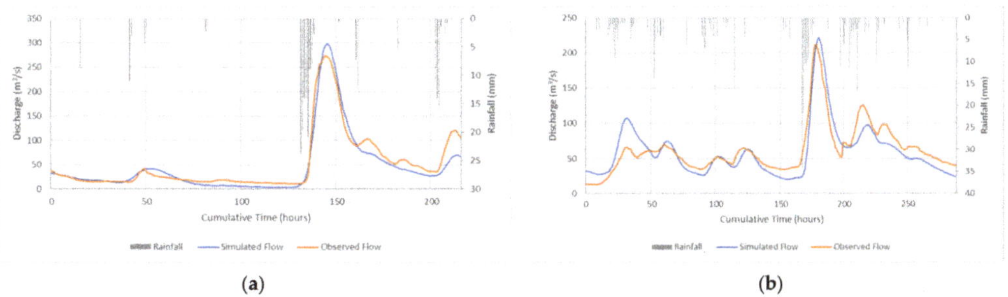

(a) (b)

Figure 2. HEC-HMS calibration at *Pitabeddara* station: (**a**) May 2018 storm event; (**b**) September 2019 storm event.

Table 3. Statistical evaluation HEC-HMS calibration and validation steps.

Title 1	Event	NSE	PBIAS	RMSE
Calibration	May 2018	0.93	−10.95%	0.3
	September 2019	0.746	−4.82%	0.5
Validation	May 2017	0.927	−8.33%	0.3

The observed and simulated hydrograph at *Pitabeddara* gauging stations for the May 2017 event is shown in Figure 3. The calibrated parameters were averaged and validated using the May 2017 event. As mentioned earlier, this event is a comparatively more severe flood event, and rainfall frequency analysis has classified this event as having a return interval of 75–100 years [49]. Due to the severity, the gauge station at *Pitabeddara* was completely flooded during the event's peak. The peak discharge of the flood was estimated by flood marks. Thus, there remains considerable uncertainty in the peak discharge of this event. However, the simulated flow using averaged parameters showed the goodness of fit with the observed flow with an NSE value of 0.927, a PBIAS of −8.33%, and an RMSE value of 0.3. As with calibration, the PBIAS indicates a model overestimation bias.

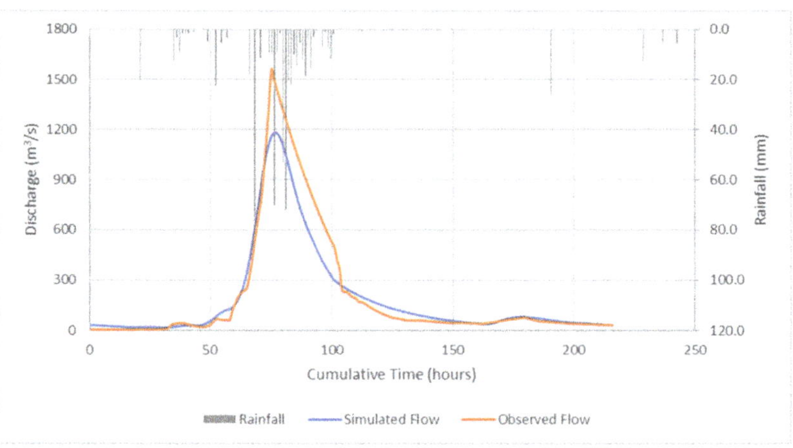

Figure 3. HEC-HMS Validation using May 2017 flood event at *Pitabeddara*.

3.2. Hydraulic Model Simulation

Initial simulation runs identified that the model could not correctly identify the flow path of the mainstream in specific regions due to the coarser 30 m DEM used. As a result, the elevation file was modified to indicate the river's correct flow path. A study conducted for flood simulation in Kabul using iRIC, Shokory et al., [28] modified the DEM similarly as the flood would not pass over the original DEM. Initially, the model was applied to a smaller region, including Matara and the region immediately north of the city. This overestimated the city's flood as the model failed to account for flooding that occurred above this region. Thus, the simulation extent was extended further upstream up to *Akuressa* for more accurate simulation.

Six inlets were located within the simulation extent (Figure 4a). The hydrographs in Figure 4b, which were obtained using the HEC-HMS model were set as the boundary conditions for Nays2DFlood at each inlet. The highest contribution was seen from inlet 1, the main river, and included flow generated from a larger proportion of the upper catchment. Inlet 2, the second-largest contributor, represents a tributary of the *Nilwala* river, *Kirama Oya*.

Paddy cultivation represents the primary land use of the region surrounding Matara city. The 30 m DEM could not recognize the complicated network of interconnected minor canals that characterize this region. The roughness coefficient was taken as 0.15 for forested areas and 0.09 [46] for other areas based on land cover and by reviewing past studies and literature [41]. Moreover, a section of Matara city was incorrectly simulated as being flooded when in fact flooding did not occur. We attribute this to faults in the DEM, that was unable to detect flood control structures such as bunds on either side of the river due to topographical changes that might have occurred owing to its production and low resolution.

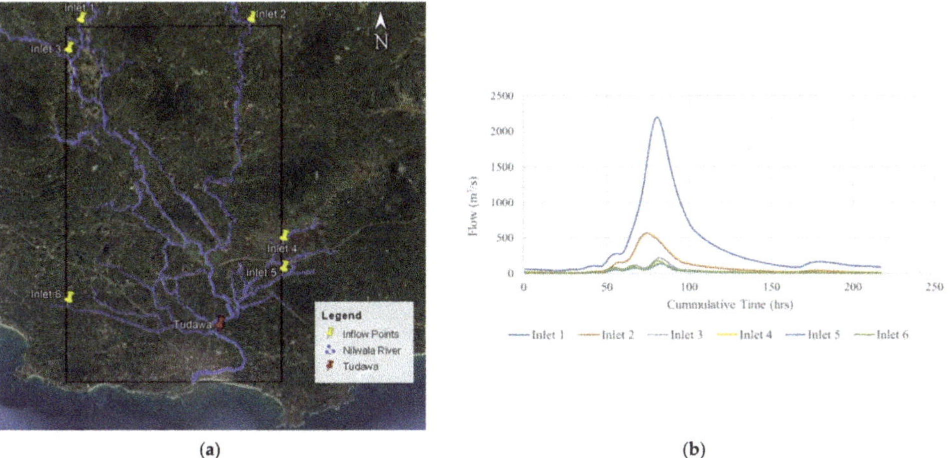

(**a**) (**b**)

Figure 4. (**a**) 2D flood modeling extent and inflow locations. (**b**) Input hydrographs for flood modeling.

The extent of flooding was verified using a survey of the flood extent carried out by the Department of Irrigation for the 2017 May flood event (Figure 5). The survey was carried out by considering flood marks and accounts of the residents in the months shortly after the flood event. An accuracy assessment (Table 4) using 10,000 random points was carried out to compare the simulated flood extent with the surveyed extent, with a maximum simulated depth greater than 0.5 m being considered as a flood, based on the classification of Hamdan et al., [50]. The producer's accuracy details the number of reference points (i.e., surveyed points) classified correctly to the total number of reference points for that particular class. The user's accuracy represents the correctly simulated points to the total number of points simulated for its respective class. The results of the assessment indicated an overall accuracy of 81.5%. The relatively lower producer's accuracy for inundated class (70.1%) when compared with the higher user's accuracy (81.5%) indicates a higher error of observed inundated points being simulated as unaffected.

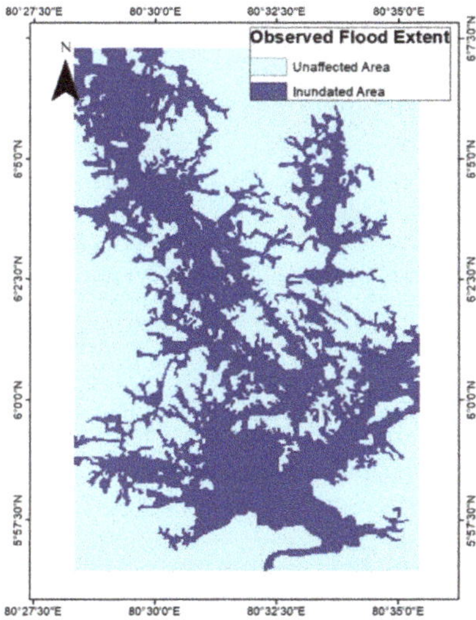

Figure 5. The surveyed flood extent of May 2017 flood event (Source: Department of Irrigation, Sri Lanka).

Table 4. Accuracy assessment of flood extent simulation using iRIC.

Type	Inundated/Unaffected	Accuracy (%)
Producer's Accuracy	Inundated	70.10
	Unaffected	88.72
User's Accuracy	Inundated	79.76
	Unaffected	82.40
Overall Accuracy		81.50

The simulated flood depth accuracy was verified using observed water levels at *Tudawa* station, located north of Matara city. This resulted in; NSE = 0.94, PBIAS = −4.28, RMSE = 0.18 and R^2 = 0.95 (Figure 6).

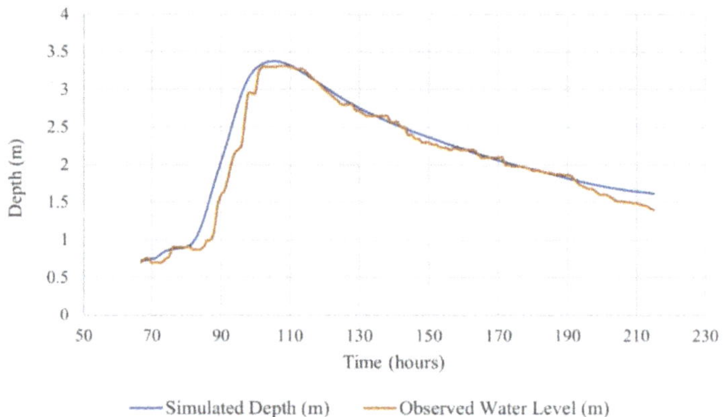

Figure 6. Comparison of observed water level and simulated flood depth of May 2017 flood at *Tudawa*.

In Figure 7a, depths over 7 m were observed towards the northern aspect of the simulation extent close to the main river's inflow points and its tributary *Kirama Oya* (i.e., inlets 1 and 2, from Figure 4). There remains uncertainty in flood depth in the upper regions as the model fails to consider any inundation that may have occurred upstream of the simulation extent which is modelled as purely one-dimensional. Furthermore, the conservation of momentum flow and the return flow of water at the linkage sites were neglected [18]. However, when considering the region closer to Matara, a maximum depth of 5 m was simulated. Comparing the velocity magnitude map (Figure 7b) with the background image affirms that the model could correctly identify the river's flow path.

Shokory et al., [28] studied the influence of rainfall and streamflow on flooding. The study found that rainfall did not significantly impact flooding in the Kabul case study, with the difference between the two resulting in the formation of "puddles". The *Nilwala* basin showed similar characteristics as the flood occurs mainly due to heavy precipitation in the upper catchment [10]. This can be clearly understood when comparing the input hydrographs generated using HEC-HMS for the May 2017 event (Figure 4). In addition, a rainfall gauging station with the capability to measure hourly precipitation within the simulated extent was unavailable. Due to these reasons, the rainfall within the basin was omitted from the simulation.

As this study seeks to develop a tool that could be applied for real-time flood forecasting, the time required to simulate needs to be considered. The grid size of the generated mesh will affect the simulation time where a grid with higher resolution will produce more accurate results but will require a significantly longer time to complete the simulation. Thus, an 80 m grid size was selected to produce results faster with acceptable accuracy.

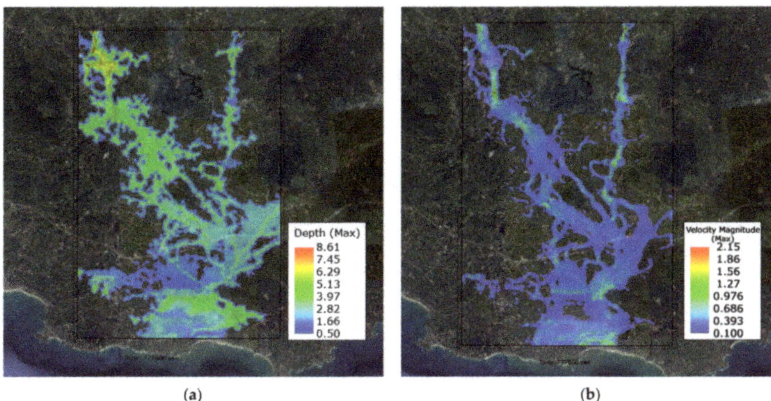

(a) (b)

Figure 7. iRIC simulation results of 2017 May flood event: (**a**) maximum flood depth and (**b**) maximum flood velocity at *Nilwala* river downstream area (Matara City and surrounding area).

During the May 2017 flood event, it took approximately 36 h for the flood peak to reach *Tudawa*. The average time taken to run the hydraulic model was close to 5 h, but this could be further reduced by using a machine with greater computational power. However, the model did simulate within an acceptable period, with sufficient lead time for taking necessary precautions based on the generated flood risk maps.

A flood risk map (Figure 8) was generated based on the May 2017 flood event. Regions were classified as high, medium and low risk according to the flood depth [50]. Such maps generated for different return periods can aid in the design and planning of infrastructure, flood control structures, and provide guidance in community resiliency development projects. In addition, maps generated in real-time during a storm event can be used to issue flood warnings and provide specific instructions for evacuation where needed.

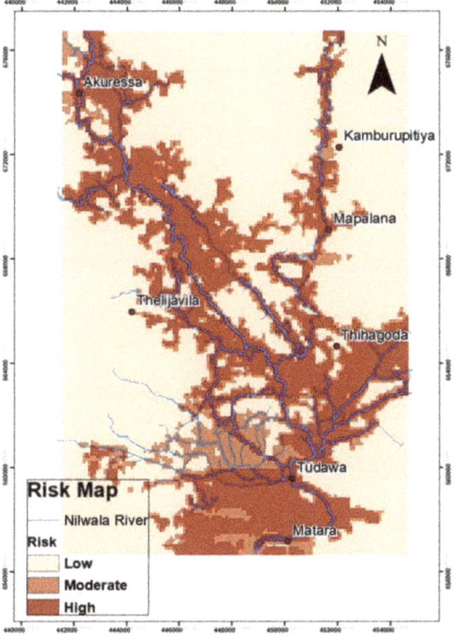

Figure 8. Flood risk map of *Nilwala* River downstream (Matara City and surrounding area) for May 2017 flood event.

4. Conclusions

The *Nilwala* river basin's upper region was modelled in HEC-HMS using the Snyder unit hydrograph method to simulate transformation, while the Muskingum–Cunge routing model, deficit and constant loss model, and recession baseflow model were used to model routing, loss and baseflow, respectively. The model was calibrated and validated using three flood events: May 2018 and September 2019 events for calibration and the May 2017 event for validation. The model performed well considering the Nash–Sutcliffe efficiency, percent bias, and root mean square error. Simulated stream hydrographs from HEC-HMS were externally coupled with iRIC to simulate inundation in Matara for the 2017 May flood event. The neglection of upstream flooding, momentum transfer and return flow at the linkage site presented major limitations of the model. The accuracy of the DEM used will greatly impact the performance of iRIC. However, the extent of flooding was estimated, with an overall accuracy of 81.50%. The accuracy of simulated flood depth was verified using water levels at the *Tudawa* gauging station which resulted in an NSE of 0.94, a PBIAS of −4.28, an RMSE of 0.18 and an R^2 of 0.95. Based on these values, it can be concluded that the 1D–2D coupled model can accurately identify the extent of flooding in addition to the flood depth. Nevertheless, the model's accuracy should be further improved by simulating more events with available observed flood data. Moreover, the model should be tested in the field using real-time flood events to evaluate its feasibility as a tool for flood forecasting and early warning systems. Accordingly, a flood risk map was developed for the May 2017 flood event. Further studies should be undertaken to develop such maps for storms with different return periods and for various climate change scenarios.

Author Contributions: Conceptualization, M.H.J.P.G. and K.S.; methodology, L.D. and M.R., M.K.N.K.; software, L.D. and M.R.; validation, M.H.J.P.G., M.K.N.K. and M.R.; data curation, T.J.M.; writing—original draft preparation, L.D.; writing—review and editing, M.H.J.P.G. and K.S.; supervision, M.H.J.P.G. and M.K.N.K. All authors have read and agreed to the published version of the manuscript.

Funding: This research received no external funding.

Conflicts of Interest: The authors declare no conflict of interest.

References

1. IPCC. *Climate Change 2013—The Physical Science Basis. Contribution of Group I to the Fifth Assessment report of the Intergovernmental panel on Climate Change*; Stocker, T., Plattner, G., Tignor, M., Allen, S., Boschung, J., Nauels, A., Xia, Y., Bex, V., Midgley, P., Eds.; Cambridge University Press: Cambridge, UK; New York, NY, USA, 2013.
2. Eckstein, D.; Kunzel, V.; Schafer, L.; Winges, M. *Global Climate Risk Index 2020*; Germanwatch: Berlin, Germany, 2019.
3. Karunathilaka, K.L.A.A.; Dabare, H.K.V.; Nandalal, K.D.W. Changes in Rainfall in Sri Lanka during 1966–2015. *Eng. J. Inst. Eng. Sri Lanka* **2017**, *50*, 39–48. [CrossRef]
4. Naveendrakumar, G.; Vithanage, M.; Kwon, H.H.; Iqbal, M.C.M.; Pathmarajah, S.; Obeysekera, J. Five decadal trends in averages and extremes of rainfall and temperature in Sri Lanka. *Adv. Meteorol.* **2018**, *2018*, 4217917. [CrossRef]
5. Gunarathna, M.H.J.P.; Kumari, M.K.N. Rainfall trends in Anuradhapura: Rainfall analysis for agricultural planning. *Rajarata Univ. J.* **2013**, *1*, 38–44.
6. Smith, K.; Ward, R. *Floods: Physical Processes and Human Impacts*; Wiley: Hoboken, NJ, USA, 1998; ISBN 978-0-471-95248-0.
7. Herath, S. *Flood Characteristics and Mitigation Issues Flood Trends around the world. UNU—Japan-Keio Joint Training Workshop on "Water Governance"*; United Nations University Centre: Tokyo, Japan, 2002.
8. Ministry of Disaster Management; Ministry of National Policies and Economic Affairs. *Sri Lanka Rapid Post Disaster Needs Assessment: Floods and Landslides*; Ministry of Disaster Management: Colombo, Sri Lanka; Ministry of National Policies and Economic Affairs: Colombo, Sri Lanka, 2017.
9. Dayaratne, S.T.; Perera, B.J.C. Regionalisation of impervious area parameters of urban drainage models. *Urban Water J.* **2008**, *5*, 231–246. [CrossRef]
10. Weerasinghe, K.D.N.; Elkaduwa, W.K.B.; Panabokke, C.R. Agro-climatic risk and irrigation need of the Nilwala Basin of southern Sri Lanka. In Proceedings of the Irrigation Drainage in the New Millenium, Fort Collins, CO, USA, 20–24 June 2000; pp. 455–469.
11. Jonkman, S.N.; Vrijling, J.K. Loss of life due to floods. *J. Flood Risk Manag.* **2008**, *1*, 43–56. [CrossRef]
12. Vojinovic, Z.; Tutulic, D. On the use of 1d and coupled 1d-2d modelling approaches for assessment of flood damage in urban areas. *Urban Water J.* **2009**, *6*, 183–199. [CrossRef]
13. Timbadiya, P.V.; Patel, P.L.; Porey, P.D. A 1D–2D Coupled Hydrodynamic Model for River Flood Prediction in a Coastal Urban Floodplain. *J. Hydrol. Eng.* **2015**, *20*, 05014017. [CrossRef]

14. Morales-Hernández, M.; Petaccia, G.; Brufau, P.; García-Navarro, P. Conservative 1D-2D coupled numerical strategies applied to river flooding: The Tiber (Rome). *Appl. Math. Model.* **2016**, *40*, 2087–2105. [CrossRef]
15. Samuels, D. Cross-section locations in 1-D models. In Proceedings of the 2nd International Conference on River Flood Hydraulics, Wallingford, UK, 17–20 September 1990; White, W., Watts, J., Eds.; Wiley: Chichester, UK, 1990; pp. 339–350.
16. Hunter, N.M.; Bates, P.D.; Horritt, M.S.; Wilson, M.D. Simple spatially-distributed models for predicting flood inundation: A review. *Geomorphology* **2007**, *90*, 208–225. [CrossRef]
17. Anees, M.; Abdullah, K.; Nordin, M.N.; Rahman, N.N.N.; Syakir, M.; Kadir, O. One- and Two-Dimensional Hydrological Modelling and Their Uncertainties. In *Flood Risk Management*; Hromadka, T., Rao, P., Eds.; IntechOpen: London, UK, 2017; p. 324.
18. Teng, J.; Jakeman, A.J.; Vaze, J.; Croke, B.F.W.; Dutta, D.; Kim, S. Flood inundation modelling: A review of methods, recent advances and uncertainty analysis. *Environ. Model. Softw.* **2017**, *90*, 201–216. [CrossRef]
19. Rai, P.K.; Dhanya, C.T.; Chahar, B.R. Coupling of 1D models (SWAT and SWMM) with 2D model (iRIC) for mapping inundation in Brahmani and Baitarani river delta. *Nat. Hazards* **2018**, *92*, 1821–1840. [CrossRef]
20. Lea, D.; Yeonsu, K.; Hyunuk, A. Case study of HEC-RAS 1D-2D coupling simulation: 2002 Baeksan flood event in Korea. *Water* **2019**, *11*, 2048. [CrossRef]
21. Knebl, M.R.; Yang, Z.L.; Hutchison, K.; Maidment, D.R. Regional scale flood modeling using NEXRAD rainfall, GIS, and HEC-HMS/RAS: A case study for the San Antonio River Basin Summer 2002 storm event. *J. Environ. Manag.* **2005**, *75*, 325–336. [CrossRef] [PubMed]
22. Zhang, W.; Zhang, X.; Liu, Y.; Tang, W.; Xu, J.; Fu, Z. Assessment of flood inundation by coupled 1d/2d hydrodynamic modeling: A case study in mountainous watersheds along the coast of southeast China. *Water* **2020**, *12*, 822. [CrossRef]
23. Halwatura, D.; Najim, M.M.M. Application of the HEC-HMS model for runoff simulation in a tropical catchment. *Environ. Model. Softw.* **2013**, *46*, 155–162. [CrossRef]
24. Gumindoga, W.; Rwasoka, D.T.; Nhapi, I.; Dube, T. Ungauged runoff simulation in Upper Manyame Catchment, Zimbabwe: Application of the HEC-HMS model. *Phys. Chem. Earth* **2016**, *100*, 371–382. [CrossRef]
25. Tassew, B.; Belete, M.; Miegel, K. Application of HEC-HMS model for flow simulation in the Lake Tana Basin: The case of Gigel Abay Catchment, Upper Blue Nile Basin, Ethiopia. *Hydrology* **2019**, *6*, 21. [CrossRef]
26. Wongsa, S. Simulation of Thailand Flood 2011. *Int. J. Eng. Technol.* **2014**, *6*, 452–458. [CrossRef]
27. Jamrussri, S.; Toda, Y. Simulating past severe flood events to evaluate the effectiveness of nonstructural flood countermeasures in the upper Chao Phraya River Basin, Thailand. *J. Hydrol. Reg. Stud.* **2017**, *10*, 82–94. [CrossRef]
28. Shokory, J.A.N.; Tsutsumi, J.I.G.; Sakai, K. Flood Modeling and Simulation using iRIC: A Case Study of Kabul City. *E3S Web Conf.* **2016**, *7*, 5–11. [CrossRef]
29. Gopinatha, S.; Nagarajanb, N. Hydrodynamic simulation of urban stormwater drain (Delhi city, India) using iRIC model. *J. Appl. Res. Technol.* **2018**, *16*, 374–381.
30. Nelson, J.M.; Shimizu, Y.; Abe, T.; Asahi, K.; Gamou, M.; Inoue, T.; Iwasaki, T.; Kakinuma, T.; Kawamura, S.; Kimura, I.; et al. The international river interface cooperative: Public domain flow and morphodynamics software for education and applications. *Adv. Water Resour.* **2016**, *93*, 62–74. [CrossRef]
31. Atkins. Atkins CRP-DBIP Computational framework specific to Nilwala Ganga basin. In *Climate Resilience Improvement Project (CRIP)*; Department of Irrigation: Colombo, Sri Lanka, 2017.
32. Jayawardana, J.M.C.K.; Jayathunga, T.R.; Edirisinghe, E.A. Water quality of Nilwala River, Sri Lanka in relation to land use practices. *Sri Lanka J. Aquat. Sci.* **2016**, *21*, 77–94. [CrossRef]
33. De Silva, M.P.; Karunatileka, R.; Thiemann, W. Study of some physicochemical properties of Nilwala River waters in Southern Sri Lanka with special reference to effluents resulting from anthropogenic activities. *J. Environ. Sci. Health Part A Environ. Sci. Eng.* **1988**, *23*, 381–398. [CrossRef]
34. Scharffenberg, B.; Bartles, M.; Brauer, T.; Fleming, M.; Karlovits, G. *Hydrologic Modeling System HEC-HMS—User's Manual*; USACE: Washington, DC, USA, 2018.
35. USACE. *Hydrologic Modeling System, HEC-HMS Applications Guide*; USACE: Washington, DC, USA, 2017.
36. Thiessen, A. Precipitation averages for large areas. *Mon. Weather Rev.* **1911**, *39*, 1082–1089. [CrossRef]
37. Snyder, F.F. Synthetic unit-graphs. *Eos Trans. Am. Geophys. Union* **1938**, *19*, 447–454. [CrossRef]
38. Te Chow, V.; Maidment, D.R.; Mays, L.W. *Applied Hydrology*, 2nd ed.; McGraw Hill Book Co.: Singapore, 1988.
39. Cunge, J.A. On the subject of a flood propagation computation method (Muskingum method). *J. Hydraul. Res.* **1969**, *7*, 205–230. [CrossRef]
40. Chow, V.T. *Open Channel Hydraulics*; McGraw-Hill: New York, NY, USA, 1959.
41. Arcement, G.J.; Schneider, V.R. Guide for Selecting Manning's Roughness Coefficients for Natural Channels and Flood Plains. *Geol. Surv. Water-Supply Pap. 2339* **1993**, *42*, 350–355.
42. Shimizu, Y.; Inoue, T.; Suzuki, E.; Kawamura, S.; Iwasaki, T.; Hamaki, M.; Omura, K.; Kakegawa, E.; Yoshida, T. *iRIC Software Nays2DFlood Solver Manual*; The International River Interface Cooperative (iRIC): Hokkaido, Japan, 2015.
43. Nash, J.; Sutcliffe, J. River flow forecasting through conceptual models, part 1—A discussion of principles. *J. Hydrol.* **1970**, *10*, 282–290. [CrossRef]
44. Gupta, H.V.; Sorooshian, S.; Yapo, P.O. Status of Automatic Calibration for Hydrologic Models: Comparison With Multilevel Expert Calibration. *J. Hydrol. Eng.* **1999**, *4*, 135–143. [CrossRef]

45. Ritter, A.; Muñoz-Carpena, R. Performance evaluation of hydrological models: Statistical significance for reducing subjectivity in goodness-of-fit assessments. *J. Hydrol.* **2013**, *480*, 33–45. [CrossRef]
46. Ratnayake, U.; Sachindra, D.A.; Nandalal, K.D.W. Rainfall Forecasting for Flood Prediction in the Nilwala Basin. In Proceedings of the International Conference on Sustainable Built Environment (ICSBE-2010), Kandy, Sri Lanka, 13–14 December 2010.
47. Feldman, A. *Hydrologic Modeling System Technical Reference Manual*; USACE: Washington, DC, USA, 2000.
48. Moriasi, D.N.; Arnold, J.G.; Van Liew, M.W.; Bingner, R.L.; Harmel, R.D.; Veith, T.L. Model Evaluation Guidelines for Systematic Quantification of Accuracy in Watershed Simulations. *Am. Soc. Agric. Biol. Eng.* **2007**, *50*, 885–900. [CrossRef]
49. Irrigation Department. *Inundation Maps of the Nilwala River Flood in May 2017*; Irrigation Department: Colombo, Sri Lanka, 2017.
50. Hamdan, A.N.A.; Abbas, A.A.; Najm, A.T. Flood hazard analysis of proposed regulator on Shatt Al-Arab River. *Hydrology* **2019**, *6*, 80. [CrossRef]

MDPI

Article

Flood Mapping from Dam Break Due to Peak Inflow: A Coupled Rainfall–Runoff and Hydraulic Models Approach

Mihretab G. Tedla [1], Younghyun Cho [1,2,*] and Kyungsoo Jun [1]

[1] Graduate School of Water Resources, Sungkyunkwan University, 2066 Seobu-ro, Jangan-gu, Suwon 440-746, Korea; mihretab1@g.skku.edu (M.G.T.); ksjun@skku.edu (K.J.)

[2] K-water Research Institute, Korea Water Resources Corporation (K-water), 1689 beon-gil 125, Yuseong-daero, Yuseong-gu, Daejeon 34045, Korea

* Correspondence: yhcho@kwater.or.kr; Tel.: +82-42-870-7476

Abstract: In this study, we conducted flood mapping of a hypothetical dam break by coupling the Hydrologic Engineering Center's Hydrologic Modeling System (HEC-HMS) and River Analysis System (HEC-RAS) models under different return periods of flood inflow. This study is presented as a case study on the Kesem embankment dam in Ethiopia. Hourly hydrological and meteorological data and high-resolution land surface datasets were used to simulate the design floods for piping dam failure with empirical dam breach methods. Based on the extreme inflows and the dam physical characteristics, the dam failure was simulated by a two-dimensional, unsteady flow hydrodynamic model. As a result, the dam will remain safe for up to 50-year return-period inflows, but it breaks for 100- and 200-year return periods and floods the downstream area. For the 100-year peak inflow, a 208 km^2 area will be inundated by a maximum depth of 20 m and for a maximum duration of 46 h. The 200-year inflow will inundate a 240 km^2 area with a maximum depth of 31 m for a maximum duration of 93 h. The 2D flood map provides satisfactory spatial and temporal resolution of the inundated area for evaluation of the affected facilities.

Keywords: design floods; HEC-HMS; HEC-RAS; dam break; unsteady; flood mapping; Kesem

Citation: Tedla, M.G.; Cho, Y.; Jun, K. Flood Mapping from Dam Break Due to Peak Inflow: A Coupled Rainfall–Runoff and Hydraulic Models Approach. *Hydrology* **2021**, *8*, 89. https://doi.org/10.3390/hydrology8020089

Academic Editors: Marina Iosub and Andrei Enea

Received: 28 April 2021
Accepted: 31 May 2021
Published: 6 June 2021

1. Introduction

Dam failures have recently had a large disastrous effect on downstream areas in many countries. Between 2000 and 2009, more than 200 notable dam failures occurred worldwide [1,2], such as Oroville, USA, in 2017 (dam failure) [3]; Brumadinho, Brazil, in 2019 (dam break) [4]; Xepian Xe Nam Noy Dam, Laos, in 2018 (dam break) [5]; and Hidroituango, Colombia, in 2018 (dam failure) [6]. In the east Africa region, water burst through the banks of the Patel Dam in Kenya's Rift Valley, washing away almost an entire village [7]. Public concern has resulted in an increasing focus on dam safety and has imposed responsibility on decision makers. Dam-break-induced disasters may occur more frequently due to infrastructure ageing. Emergency planning, as a non-structural measure to minimize flood impacts, plays an important role in crisis management. One of the keys to preventing and reducing losses from flood disasters is obtaining reliable information about the flood risk through flood inundation maps [8]. If a disaster cannot be avoided, individual and social structure preparedness may help in risk reduction [9]. As the cost of hazard assessment is negligible compared with the total cost of disasters, every dam should have been analyzed for safety with an emergency plan in place [10].

Numerical flood simulation models are useful tools for quantitatively evaluating flooding, which enable researchers to qualitatively evaluate hazard and other risks caused by flooding; however, these models have limitations related to insufficient datasets [11]. Recently, due to advances in hydraulic models, research on the application of 2D models for flood risk management, including dam breaks [12,13], has been increasing. Despite the rapid development of the risk analysis of dam breaks, research is relatively lacking on integrating the impact of peak inflows, which causes dam breaks. Heavy rainfall and peak flood inflows have the potential to trigger dam failures. As a result, determining their impacts on dam failure and the associated disaster is instrumental in understanding complicated dam break problems. Moreover, the flood risk analysis, the inundation extent, and the environmental impact of dam breaks are also affected by peak flood inflows.

In this study, the Hydrologic Engineering Center's Hydrological Modeling System (HEC-HMS), which can simulate a wide range of hydro-meteorological and physical processes including flooding, and the Hydrologic Engineering Center's River Analysis System (HEC-RAS) 2D model were used to evaluate the impact of peak inflows on dam break and flood inundation. HEC-GeoHMS is typically used to import all the required inputs into HEC-HMS [14]. The watershed is physically represented with the HEC-HMS semi-distributed basin model [15]. Various hydrologic elements are connected in a dendritic network to simulate runoff processes [16,17]. In this study, the inflow hydrographs for different return periods were generated with hourly precipitation and streamflow data using the curve number method. RAS Mapper, an extension of HEC-RAS, has a set of procedures, tools, and utilities to explicitly display the results [18]. HEC-RAS 2D has a GIS interface and applies the finite method to solve unsteady flow equations that describe the two dimensions of the river and floodplain [19,20]. A detailed animation showing flood wave progression in multiple directions on a local scale is best-represented using a 2D model.

The novel objective of this study was to develop a flood inundation map by integrating hydrological and hydraulic models for a hypothetical dam break. The peak inflow for various return periods was used to initiate the dam break. The inputs derived from the physical condition, material properties, dimensions of the dam, and maximum bounding breach width and height were used in the empirical breach method. A stepwise approach was implemented to execute the model. The first step was developing a physically based hydrological model to simulate the designed rainfall for flood discharges with HEC-HMS after calibrating and validating using historical storm events. The output of this step was used to predict flood volume, peak rates, and the runoff hydrograph. The next step was to develop a hydraulic model of the watershed for a piping dam break with an HEC-RAS model. The last step was to simulate the flood hydrograph and analyze the inundation map of the return periods using the 2D flood map.

This study is presented as a case study of the Kesem embankment dam. The aim was to understand the interaction of hydro-meteorological, topographic, and physical dam settings to determine dam-break-associated flooding. The Afar region, located in the East African Rift System, has been affected by progressive stages in the development of the largest active rift in the world [21]. Given its volcanically and seismically active setting, the Kesem dam is subjected to high seismicity due to regular strain release occurrences [22]. Previously, dam breach simulation and analysis of the Kesem dam were conducted by combining the physical breaching model HR BREACH, a simple geotechnical failure model, to simulate any breaches in the embankments, with HEC-RAS 1-D, to perform the flood routing [23]. In our method, peak inflows drive the dam break in the piping dam failure method, demonstrating the inundation area on a 2D flood map.

Study Area

Kesem River is a tributary to the Awash River Basin, which is the most-used among the twelve basins in Ethiopia (Figure 1a). Annual rainfall is received in the four rainy months of June, July, August, and September, and most of the river courses become full and flood their surroundings during these months [11]. The Kesem dam was built in 2015 by the then Ethiopian Water Work Construction Enterprise (WWCE), mainly for sugarcane irrigation, with the capacity for developing 20,000 ha of land. The dam is a rockfill dam, categorized as a large dam, possessing an embankment volume of 3.15 hm^3 and a crest length of 685 m. The capacity of the reservoir is estimated at 770 hm^3, of which 360–480 hm^3 (47–62%)

is assumed to be live storage, and the spillway capacity is about 6180 m^3/s. The dam is located 237 km to the northeast of Addis Ababa at 800 m above sea level.

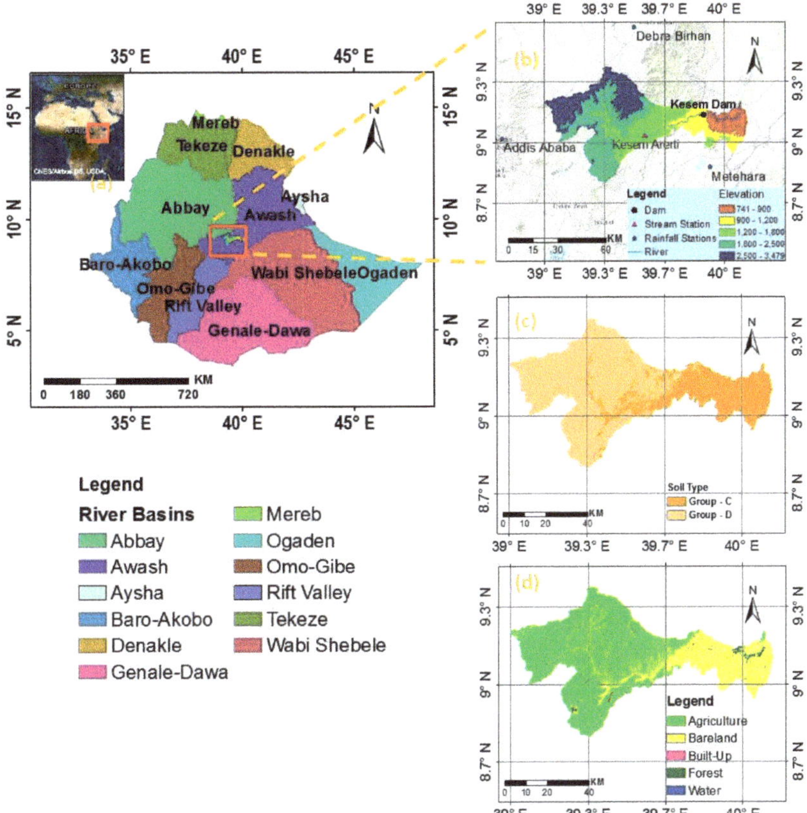

Figure 1. Locations and characteristics of the study area: (**a**) Ethiopian river basins and the Kesem watershed in the Awash river basin; (**b**) elevation, Hydro-Met observation stations, and Kesem dam location; (**c**) hydrologic soil group; and (**d**) land-use map of the Kesem watershed.

2. Materials and Methods

We combined a rainfall–runoff and a hydro-dynamic model to map floods due to dam break. High spatio-temporal resolution land surface and observed hydro-climatic datasets were employed for simulating the probabilistic rainfall events of various return periods that trigger piping dam failure. We also investigated the performance of the coupled rainfall–runoff and hydraulic models in flood hazard and inundation extent analysis. The topographic data (i.e., DEM, soil, land use, and terrain data) and the hydro-meteorological data used in this study are presented in the Data Acquisition section. Figure 2 summarizes the methodological workflow followed in this study.

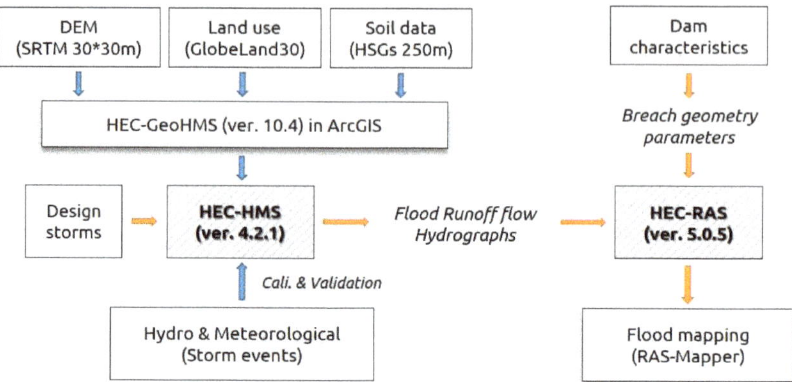

Figure 2. Overall procedures of our modeling work.

2.1. Data Acquisition

2.1.1. Land Surface Data

The global data for a digital elevation model (DEM) with a 30 m × 30 m resolution were acquired from the Shuttle Radar Topographic Mission (SRTM) of the United States Geological Survey (USGS) to extract stream networks and prepare inputs to the hydrologic model. ArcGIS (version 10.4) was used to prepare the spatial datasets for use in the HEC-HMS 4.2.1 [24] model. Shapefiles collected from the Ministry of Water Irrigation and Energy (MoWIE), Ethiopia, reservoir geometry in the form of control levels, and the revised elevations–area–capacity curve from a dam design report [25] were used to determine the breaching simulation at different water levels.

Although a wide range of soil types exists in the Awash river basin, two main soil classes, shown in Figure 1c, can be distinguished in the Kesem sub-basin, based on the global hydrologic soil groups (HYSOGs250m) for curve number-based runoff modeling [26]. The most abundant hydrologic soil group (HSG) is D soils, which have the highest runoff potential and very low infiltration rates when thoroughly wetted. They mainly consist of clay soils with a high swelling potential, soils with a permanent high-water table, soils with a claypan or clay layer at or near the surface, and shallow soils over nearly impervious material. The other HSG that exists in the study area is Group C. Group C has low infiltration rates when thoroughly wetted and mainly consists of soils with a layer that impedes downward movement of water and soils with a moderately fine structure [27].

For the land-use classification, the GlobeLand30 dataset developed by China, which has a global surface coverage dataset with a 30 m resolution, was acquired from the National Geographic Information Resource Directory Service [28]. These data contain richer and more detailed information on the distribution of global surface coverage, which can more accurately portray most human land-use activities and the landscape patterns. The land use was reclassified broadly into five groups: agriculture, bare land, forest, built-up, and water. Table 1 shows the major land-use areas in km^2 in the Kesem watershed.

Table 1. Land-use classifications and corresponding areas in the Kesem watershed.

Land Use	Contents	Area (km^2)
Agriculture	Land used for agriculture, horticulture and gardens	2242.7
Bare land	Land covered by natural grass, shrubs and with vegetation less than 10%	1095.9
Forest	Land covered with trees, with vegetation cover over 30%	148.8
Built-up	Lands modified by human activities	11.1
Water	Water bodies in the land area	1.4

2.1.2. Meteorological and Hydrological Data

In this study, hourly precipitation and discharge data were used for the model simulation. The rainfall data were collected from the Ethiopia National Meteorological Agency (NMA) at three observation stations. Similarly, hourly streamflow data from MOWIE were collected at the automated Kesem Arerti gauge station. The locations and description of these stations are provided in Figure 1 and Table 2. Both the precipitation and streamflow data were originally collected in 15 min intervals and were converted to hourly intervals. An automated real-time telemetry observation system was used to collect the streamflow data; the precipitation data were collected from first-class meteorological stations. In Ethiopia, only the first-class meteorological stations can collect and store high-temporal-resolution data, up to 15 min, with many climate parameters. For the hydrologic model set-up, five storm events (16–19 August 2012, 22–24 August 2013, 22–24 August 2016, 7–9 August 2014, and 18–20 August 2015) were used independently to calibrate and validate the model.

Table 2. List of selected meteorological and hydrological gauge stations.

Type	Name	Latitude (Decimal Degrees)	Longitude (Decimal Degrees)	Period (Years)
Precipitation	Addis Ababa	9.02	38.77	2012–2018
Precipitation	Metehara	8.86	39.92	2012–2018
Precipitation	Debre Birhan	9.63	39.50	2012–2018
Stream flow	Kesem Arerti	9.04	39.56	2012–2018

2.2. *Estimation of the Designed Storm*

In this study, the rainfall distributions (i.e., designed storm) for the different return periods were estimated using the Gumbel distribution. The Gumbel distribution method is best for identifying and predicting future rainfall occurrences and flood analysis [29]. The equation for fitting the Gumbel distribution using the observed series of annual peak storm events (i.e., maximum precipitation) for the return period T is

$$\mathrm{P_t} = \mathrm{P_{av}} + \mathrm{K_T}\sigma \tag{1}$$

where $\mathrm{P_t}$ denotes the precipitation depth of the T-year storm event; $\mathrm{K_T}$ is the frequency factor; and $\mathrm{P_{av}}$ and σ are the mean and the standard deviation of the annual maximum precipitation, respectively. The frequency factor $\mathrm{K_T}$ is expressed as

$$\mathrm{K_T} = -\frac{\sqrt{6}}{\pi}\left\{\lambda + \ln\left[\ln\left(\frac{T}{T-1}\right)\right]\right\} \tag{2}$$

where π is 3.14; λ is the Euler constant (0.5572); and ln is the natural logarithm.

From this, the rainfall intensity, It, is calculated by the precipitation depth of the T-year return period divided by a certain duration as

$$\mathrm{It} = \mathrm{P_t}/\mathrm{T_d} \tag{3}$$

where $\mathrm{T_d}$ is the rainfall duration.

2.3. *Hydrologic Modeling (HEC-HMS)*

The HEC-HMS rainfall–runoff model [30] was used to simulate inflows into the dam. HEC-HMS is capable of simulating runoff based on hourly to daily rainfall [24]. All the necessary processes to produce HEC-HMS projects were prepared using ArcHydro [31] and HEC-GeoHMS [32] tools with input files from the DEM, stream network, sub-basin boundaries, and connectivity of various hydrologic elements. The HEC-GeoHMS in the ArcGIS environment can create HEC-HMS projects using terrain pre-processing and basin processing tools [14,33]. Arc Hydro is a geospatial and temporal data model for water resources, which operates within ArcGIS. The attribute tables of the terrain were populated with consecutive calculations of basin slope, river length, river slope, basin centroid, centroid elevation, and centroid longest flow path under the Characteristics tool. The land use and soil map were employed to generate the CN grid map through the HEC-Geo HMS tool Generate CN grid.

In the hydrologic modeling, different methods were used to create the rainfall–runoff relationship. The soil conservation service-curve number (SCS-CN) method for losses, the SCS unit hydrograph for transformation, the monthly constant for base flow, and the Muskingum–Cunge method for determining the channel routing were used for the simulation. A curve number grid was generated with the HMS tool; the most common parameters in the flood simulation were entered to develop the HMS schematic for import into HEC-HMS. After performing all the necessary processes in HEC-GeoHMS, the HMS model was exported for hydrologic simulation.

The rainfall–runoff model parameters have to be derived from calibration against one or more observed variables, e.g., streamflow. Comparison of these parameters with modeled results provides physical meaning and acceptable ranges for their values [34]. Exact values for the parameters usually cannot be fixed in advance. Therefore, five hourly historical storm events between 2012 and 2018 at three precipitation gauge and one hydrological gauge stations were used to set up the model through calibration and validation. The model estimates flow output and compares it with the observed flow at the Kesem Arerti gauge station. Statistical measures were used to assess the model performance (i.e., the strengths and weaknesses of the model). This measures-oriented approach to model performance assessment focuses on several different aspects of the overall accuracy or skill of the streamflow model. The applied model was evaluated with a coefficient of statistics, to provide the range of information [35].

The coefficient of determination (R^2) is used to describe the degree of correlation between the simulated and measured data. It ranges from −1 to 1, where values close to 1 indicate the least error. The percent bias (PBIAS) measures the tendency of the average value. The Nash–Sutcliffe efficiency (NSE) is a normalized statistic showing the residual variance, where an NSE > 0.5 is considered to indicate a good fit. The mean absolute error (MAE) and root mean square error (RMSE) are absolute measures of fit. The equations that were employed in this study are shown below:

$$R^2 = \left\{ \sum_{i=1}^{n} (\mathrm{Oi} - \mathrm{Oav})(\mathrm{Si} - \mathrm{Sav}) / \left[\sqrt{\sum_{i=1}^{n} (\mathrm{Oi} - \mathrm{Oav})^2} \sqrt{\sum_{i=1}^{n} (\mathrm{Si} - \mathrm{Sav})^2} \right] \right\}^2, \quad (4)$$

$$\mathrm{PBIAS} = \left[\sum_{i=1}^{n} (\mathrm{Oi} - \mathrm{Si}) / \sum_{i=1}^{n} \mathrm{Oi} \right] \times 100, \quad (5)$$

$$\mathrm{NSE} = 1 - \left[\sum_{i=1}^{n} (\mathrm{Oi} - \mathrm{Si})^2 / \sum_{i=1}^{n} (\mathrm{Oi} - \mathrm{Oav})^2 \right], \quad (6)$$

$$\mathrm{MAE} = \sum_{i=1}^{n} |\mathrm{Oi} - \mathrm{Si}| / n, \quad (7)$$

$$\mathrm{RMSE} = \sqrt{\sum_{i=1}^{n} (\mathrm{Oi} - \mathrm{Si})^2 / n}, \quad (8)$$

where Oi is measured flow (m^3/s); Si is simulated flow (m^3/s); and Oav and Sav are mean measured and simulated flow (m^3/se), respectively. Afterward, the frequency method was selected to run the HEC-HMS model for specified durations under the inbuilt meteorological model in the HEC-HMS model.

2.4. Hydraulic Modeling (HEC-RAS)

HEC-RAS was designed to perform one- and two-dimensional hydraulic calculations for a full network of natural and constructed channels. HEC-RAS can be used to route an inflowing flood hydrograph through a reservoir with any of the following three methods: one-dimensional (1D) unsteady flow routing (full Saint-Venant equations), two-dimensional (2D) unsteady flow routing (full Saint-Venant equations or diffusion wave equations), or level pool routing. In general, full unsteady flow routing (1D or 2D) is more accurate for both the with- and without-breach scenarios. Because HEC-RAS solves the full Saint-Venant equations, it is well-suited for computing floods [36].

In a previous study, the Kesem dam was simulated using HR BREACH for both the overtopping and piping failure modes, employing Hershfield's technique to estimate the probable maximum flood. The result showed overtopping failure is not expected as the spillway can safely evacuate this flood [23]. For piping failure, the water seeps in at a significant rate through the dam; as the material is eroded, a large hole forms. Hence, during the piping process, erosion and head cutting begin to occur on the downstream side of the dam, which eventually widen the breach. Depending on the volume of water in the reservoir, the widening continues until the natural channel bed is reached. HEC-RAS has two types of methods available for assessing the dam breach characteristics, namely, the user-entered and the simplified physical methods.

Breach formation model parameters that need to be estimated consist of the reservoir water-surface elevation at which breach formation begins (Yf), height (Hb), average width (Bav), average side slope ratio z of the final trapezoidal breach, and the breach formation time (tf, the time needed for complete development of the breach following the initiation phase), as shown in Figure 3. For this study, the Froehlich simplified physical method regression equations [37] were used for piping dam breach simulation. The relationships for both expected values and their variances are presented using the coefficient of determination (R^2) for breach width, peak outflow, and breach formation time parameters: 0.68, 0.86, and 0.96, respectively [38]. The prediction equations for the parameters Bav, z, and tf were determined from multiple regression analysis of the assembled data. Logarithmic transformations of all dependent and independent variables were found to provide the best linear relationships [39]. As a result, the Froehlich method was found to be more appropriate for the simulation of the Kesem dam.

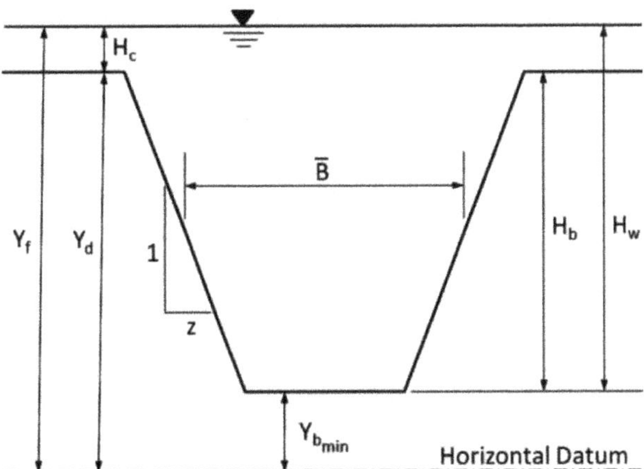

Figure 3. Dimensions of a trapezoidal dam breach approximation (Froehlich, 2008).

2.5. 2-D Flood Mapping

The modeling of a dam break flood wave is one of the most difficult unsteady flow problems to solve. Within HEC-RAS, the downstream area can be modeled as a combination of one-dimensional streams and storage areas; as a combination of one-dimensional streams, storage areas, and two-dimensional flow areas; or as a single two-dimensional flow area. Many other factors must be considered to obtain an accurate estimate of the downstream flood stages and flows. In unsteady flow model development for a dam break, stability and numerical accuracy can be improved by selecting a time step that satisfies the Courant condition. Too large a time step causes numerical diffusion (attenuation of the peak) and possible model instability, whereas too small a time step can lead to long computation times, as well as possible model instability.

The best method to estimate a computational time step for HEC-RAS is to use the Courant condition. This is especially important for dam-break flood studies. The water depth, velocity, and the water surface elevation were analyzed, and the result was displayed in HEC-RAS Mapper view. The Courant condition is shown below:

$$C = V_w \Delta T/\Delta X \leq 1 \text{ and } \Delta T \leq \Delta X/V_w \quad (9)$$

where C denotes the courant number; ΔT is time step in seconds; ΔX is the distance step in feet (average cross-section spacing or two-dimensional cell size); and V_w is the wave speed in feet per second. The flood wave speed can be calculated by

$$V_w = dQ/dA \quad (10)$$

where dQ is the change in discharge over a short time interval ($Q_2 - Q_1$); and dA is the change in cross-sectional area over a short time interval ($A_2 - A_1$).

For our applications of the Courant condition, we used the maximum average velocity from HEC-RAS and multiplied it by 1.5 to obtain a rough estimate of the flood wave speed in natural cross-sections. According to the recommendation of the HEC-RAS manual, multiplication by 1.5 can be used for practical reasons to identify the flood wave speed. The contour lines of velocity and their flow directions can also be displayed in the 2D flood map.

Afterward, a new set of layers were created and imported into the RAS mapper to visualize the outputs generated in a specified pre-processing step for performing the floodplain delineation. For the model to be accurate, small cell sizes can be used as the computational requirements are less relevant and the simulation runs quickly [12]. Previous studies have addressed the effects of mesh size and resolution on 2D flood map modeling. The coefficient of surface roughness, i.e., Manning values analysis, is required to be very large to avoid any significant impact on the model predictions [13,40]. In this study, different mesh sizes were tested considering the terrain resolution and the computational time. The maximum possible cell size, i.e., 30 m $\times$ 30 m, was employed to simulate the flood flow in a reasonable amount of computational time. In addition, a constant Manning roughness value (i.e., 0.05) was used per Chow's (1959) recommendation for vegetated flood plains [41].

Stream centerlines, cross-section cut lines, bank points, velocity points, and bounding polygons were created in ArcGIS. The software creates different bounding polygons and a spatial limit for floods based on the water surface elevation of the return periods floodplain. Finally, inundation mapping was performed in two steps: water surface generation and floodplain delineation. The water surface was created from the altitude of the water surface in the 2D flow area. Floodplain delineation was achieved using the SRTM terrain model. The flood damage and inundated infrastructure were analyzed by overlaying the Google Earth satellite map. Thus, the flood inundation boundaries and their depths were calculated. The flood inundation areas for different flood values are represented by polygons of the contour lines generated in the RAS mapper.

3. Results

3.1. Designed Storm

The designed storm was computed by applying the Gumbel distribution to each set of annual daily maximum precipitation corresponding to the return periods, as shown in Figure 4. The Gumbel distribution fairly estimates the stochasticity of the Kesem watershed, as shown on the curve below. Following the frequency result, an intensity duration frequency (IDF) curve was developed to estimate the designed storm for various return periods. The value of the IDF was converted into the depth duration frequency

(DDF) to estimate the design precipitation depth, as shown in Figure 5, and computed for the design inflow calculation.

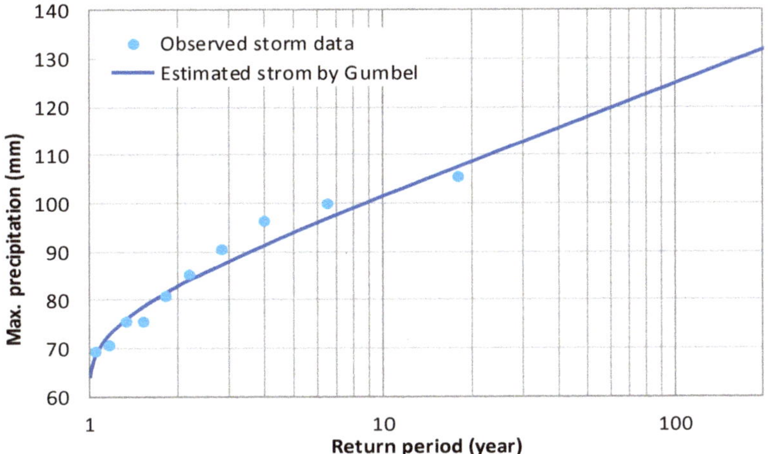

Figure 4. Frequency analysis using Gumbel frequency factors (Kesem, 2006 to 2015).

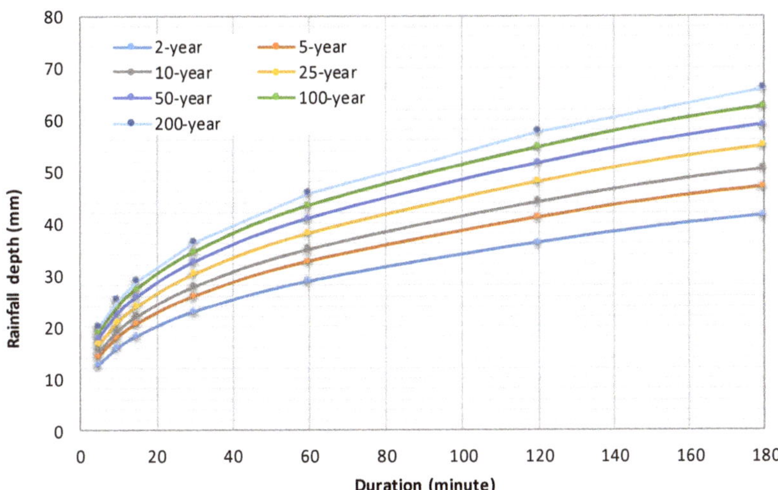

Figure 5. Depth duration frequency (DDF) for the Kesem watershed with rainfall estimation of seven return periods.

3.2. Hydrologic Simulation (Flood Hydrograph)

The HEC-HMS model was set up by simulating the historical observed data and tuning the sensitive parameters within the model. The calibration, validation, and physical properties analysis of the HEC-HMS model significantly enhanced the performance of the hydrologic model. Moreover, the base flow that was calculated from the annual minimum monthly flow of the streamflow was the crucial element in determining the peak rainfall and maintaining the hydrograph consistency during low flows. Figure 6 shows the monthly average precipitation of the Kesem watershed.

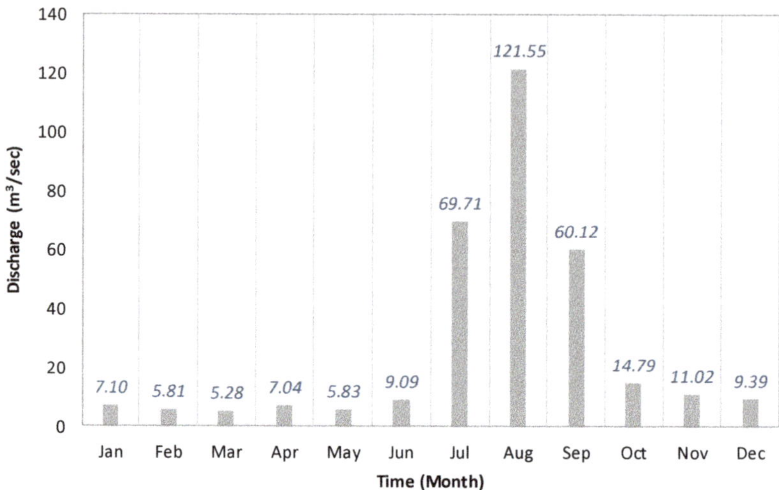

Figure 6. Annual monthly average flow at the Kesem Arerti station from 2006 to 2015.

Regardless of historical changes in the basin, the model calibration increased the efficiency of the model in making a good agreement between the simulated and observed hydrographs. Furthermore, the model was verified using peak hourly storm events in August 2014 and 2015. Both calibration and validation hydrographs are shown in Figure 7. The results of the statistical evaluations after calibration and validation are within the acceptable range, as shown in Table 3. Thus, the model was well-set and ready for the simulation of flood hydrographs using the return periods of the designed storm.

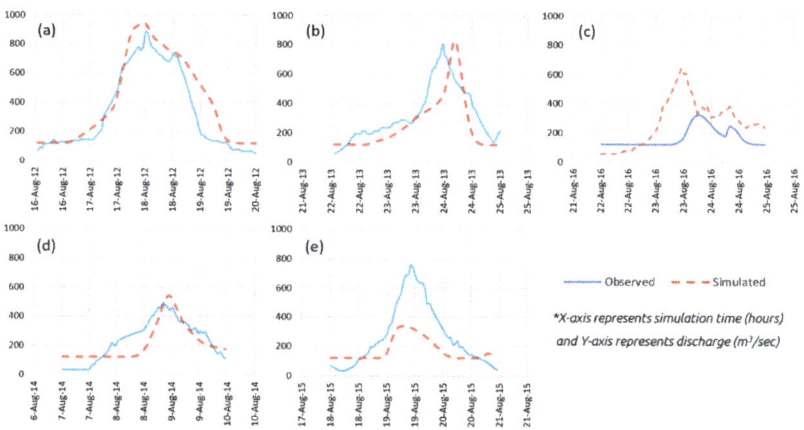

Figure 7. Model calibration and validation results: hydrographs of storm events after calibration in (**a**) 2012, (**b**) 2013, and (**c**) 2016; and validation in (**d**) 2014 and (**e**) 2015.

Table 3. Major parameter indices after calibration and validation with optimized parameter values.

Type	Storm Event Period	NSE	MAE	RMSE	R^2	PBIAS
Calibration	16–19 August 2012	0.81	75.45	117.63	0.90	22.36
	22–24 August 2013	0.50	66.51	130.25	0.66	−20.34
	22–24 August 2016	−0.30	109.79	179.80	0.21	−39.74
Validation	7–9 August 2014	0.61	21.19	86.51	0.64	−9.01
	18–20 August 2015	0.75	94.92	171.33	0.86	35.24

Afterward, the model simulated a 48 h storm event. The hydrographs from the HEC-HMS model were used as a boundary condition for the hydraulic model. Figure 8 depicts the flood hydrographs of the designed storm in the Kesem reservoir. The reservoir inflow hydrograph peak discharge for the return periods of 2, 5, 10, 25, 50, 100, and 200 years was 318.2, 968.1, 1672.6, 2866.9, 3892.3, 4997.5, and 6161.1 m^3/s, respectively. The use of hourly peak observed rainfall events for the flood estimation shows a relatively higher flood hydrograph, which increased the estimation accuracy of the Kesem watershed hydrographs.

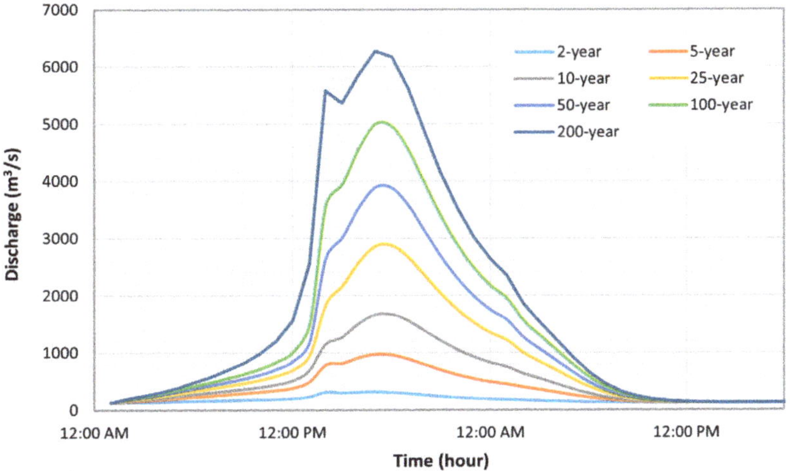

Figure 8. Reservoir flood inflow hydrographs for different return periods of the designed storm.

3.3. *Hydraulic Simulation (Flood Mapping)*

Then, HEC-RAS simulation was performed using the peak inflow hydrographs of the hydrologic model outputs as the boundary conditions for a piping dam breach. Consequently, we found the inflow up to the 50-year return period cannot result in a Kesem dam break. However, the 100- and 200-year return period inflows show dam failure and flooding of the downstream area. The dam breaching begins when the water level of the reservoir reaches an elevation of 930 m and the volume of water stored in the reservoir is 550 million m^3. The breach is estimated to begin as triangular and then stretching to reach the bottom of the breach, when the shape changes to trapezoidal. A standardized dam-breach scenario examines a range of possibilities to estimate the expected annual damages, which vary with the volume of water in the reservoir. For general planning purposes, the most useful scenario is the worst-case one in which the dam fails while the reservoir is full [42]. Therefore, in this study, the stored water in the dam and the inflow was routed to the floodplain downstream of the dam. Thus, the flood hazard maps for the 100- and 200-year return periods were generated. Figure 9 represents the model results for the return periods that show dam break and those that remain safe.

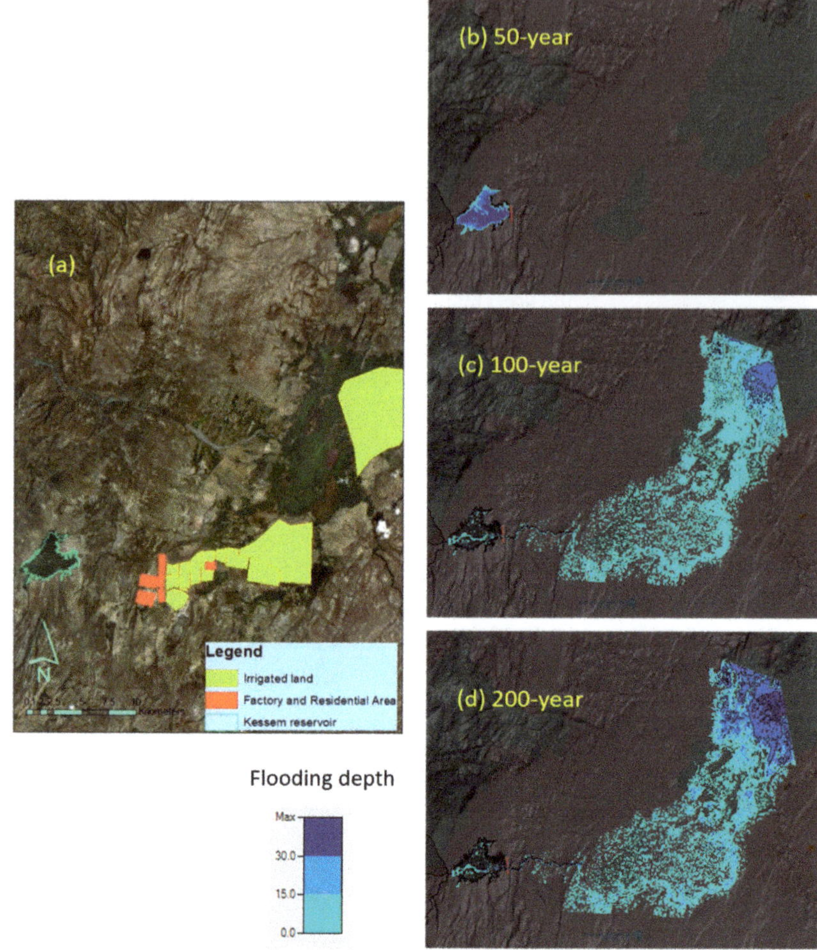

Figure 9. Hydraulic model outputs of inundation areas for various return periods: (**a**) the main infrastructure and flooded area locations; and the (**b**) 50-year, (**c**) 100-year, and (**d**) 200-year return periods.

The inundation area coverage and depth in the floodplain are displayed for both the 100- and 200-year return periods. The 2D flood simulation video in RAS mapper provides a detailed view of the flood extent, propagation, and depth. In this case, the velocity of the flooding shows variation in space and time, increasing to 15 m/s in the immediate vicinity of the dam and slowing as it flows to the wider area toward the confluence with the Awash River, as shown in Figure 10.

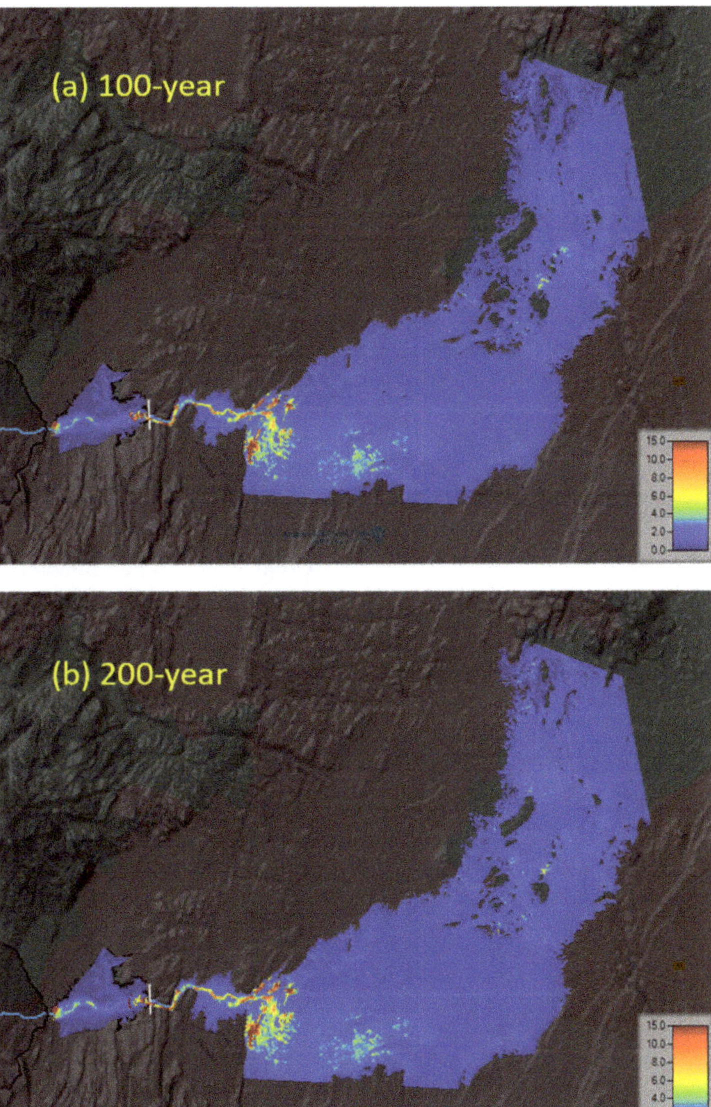

Figure 10. Maximum velocity of flood flow from the dam in the cases of (**a**) 100-year and (**b**) 200-year return period inflows.

The inundation depth and duration vary depending on the topography and proximity to the dam. Table 4 provides the main information about the flood inundation. The maximum water depth for the 100- and 200-year return periods are 20 and 31 m, respectively. The inundation time also increases to 46 and 93 h, with 10 and 39 h on average, for the 100- and 200-year return periods, respectively. The flood inundates everything in the downstream area of the dam. The devastating effect extends to the life, livelihood, and property of the community as well as the national economy, including major infrastructure such as Awash National Park, the irrigation control area, the Kesem sugar factory, and two small villages downstream of the dam.

Table 4. Main information on inundated areas from the results.

Return Period	Inundation Area (km^2)	Depth (m)		Duration (h)	
		Maximum	Mean	Maximum	Mean
100 years	208	20.06	2.04	46	10
200 years	240	31.20	2.84	93	39

4. Discussion

In previous studies, the Kesem watershed and dam flood flows were estimated using monthly to daily time-series data for flood simulation. In this study, the flood flow estimation using hourly time steps of precipitation data demonstrated a more reasonable results, as short duration peak events are included in flow calculation. That means the higher the temporal resolution of the record, the higher the accuracy of the flow estimation. Moreover, the statistical results of the simulated flow hydrograph against stream flow observations were within an acceptable range after model calibration and validation. Although the application of short-term rainfall data to the evolution of long-term trends might not be representative and conclusive, as the trend results are influenced by data uncertainty [43], the observed annual maximum storm data of the Kesem watershed well fit the estimated storm curve produced using the Gumbel method, as shown in Figure 4. In addition, the accuracy of estimation was enhanced using a longer period of observed data and the use of IDF and DDF curves with the Gumbel precipitation distribution, which were found indispensable for return periods flow simulation.

Dam breaches are often simply modeled in the shape of a trapezoid that is defined by its final height, base width, or average width and side slopes, along with the time needed for the opening to form completely [39]. In this study, the dimension and occurrence of the dam breach were estimated by Froehlich mathematical expressions for the expected values of the final width and the side slope of a trapezoidal breach, along with its formation time. Although data on historical embankment dam failures are limited and were generally not directly and accurately recorded, Froehlich collected data from 74 embankment dam failures for the determination of the breach characteristics [37]. Additionally, an investigation into the Mosul embankment dam failure with HEC-RAS embedded with piping dam breach mathematical equations showed that Froehlich is the most suitable method for estimating breach parameters from an embankment dam [44]. Therefore, the application of the Froehlich method to the Kesem dam is the most accurate method to estimate the breach dimensions and formation time.

The use of a high-spatial-resolution DEM for 2D flood map generation and inundation extent determination was essential for topographic and cross-sectional data-scarce areas. River cross-sectional dimension data at a specified interval are scarcely available for many rivers of the world. Furthermore, for flood patterns around discrete features, it is important to use a 2D model to determine the direction of flow [45]. However, terrain analysis for implementing large-scale hydraulic simulation is still a challenging and active research issue. A comparison of the detailed digital surface models (DSMs) and DEMs shows that the detailed DSMs may more accurately indicate the surface relief and the existing natural obstacles, such as vegetation, buildings, and greenhouses, enabling more realistic hydraulic simulation results [46]. Consequently, the 2D map can be enhanced by incorporating 1D simulations with the necessary topographic and cross-sectional data as well as detailed digital surface models. Additionally, in this study, the existing structures downstream the dam, except for the flood-control levee on the side of Awash River, could not be included due to a lack of data. However, the accuracy of the results can be improved with the addition of existing inline and lateral structures, such as weirs, bridges, and dykes. The effect of land use on both the watershed and the floodplain can be demonstrated to depict scenario-based flow change and the expected damage. The depth of inundation and the velocity can be synchronized to assist in practical impact-based flood forecasting and damages calculation.

Although there is no unified definition, the environmental impact of a dam break, which refers to the changes in the natural environment and ecological conditions around the reservoirs, mainly includes the natural environment, including water, soil, atmosphere, noise, and solid waste [47]. Flood maps of anticipated dam breaks cannot be accurately drawn, and it is practically impossible to exactly validate the inundation area. Flood maps can be enhanced by incorporating a validation sample specific to the flood-inundated area. We used the volume–area–elevation curve between the reservoir storage time and the breach occurrence time to validate the inundation area. The validated model could be useful for reservoir operation and dam safety management.

5. Conclusions

In this study, we coupled HEC-HMS and HEC-RAS for application to the flood mapping of a Kesem dam break in Ethiopia, with the results showing a satisfactory result for both the rainfall–runoff and hydraulic simulations. The use of 2D flood mapping, particularly in data-scarce areas, greatly contributed to the identification of risk zones and the extent of the hazard. From this study, the following conclusions were drawn:

1. The use of high-temporal-resolution hydro-meteorological data (i.e., precipitation and streamflow), for flow estimation, and high-spatial-resolution topographic data (i.e., DEM, land use, and soil), for flood inundation mapping, performs well.
2. The flood hydrographs produced from event-based runoff estimation by the SCS curve number method displayed suitable results for application to peak storm events.
3. The Kesem dam, from the empirical dam break simulations, shows possible failure for the 100- and 200-year return period inflows, whereas the dam remains safe for inflows up to the 50-year return period. This may indicate user-defined breaching, with the HEC-RAS models determining the dam breach time and dimension depending on the breach trigger inputs, such as inflow boundary conditions.
4. The 2D flood map provides a satisfactory spatial and temporal resolution result for the inundated area. Furthermore, 2D flood mapping can be used to better understand the flood inundation extent of any type of flood.

Author Contributions: Conceptualization, methodology, investigation, resources, data curation, writing original draft preparation, validation, software and formal analysis: M.G.T.; writing—review and editing, supervision and project administration: Y.C.; conceptualization and funding acquisition: K.J. All authors have read and agreed to the published version of the manuscript.

Funding: This research was done under the scholarship funded by the Korea International Cooperation Agency (KOICA).

Data Availability Statement: The data presented in this study are available on request from the first author. The data are not publicly available due to restrictions from the raw data providers.

Acknowledgments: We would like to acknowledge to all the support from the KOICA and the graduate school of water resources at Sungkyunkwan University. The author thanks K-water and the USACE for training in the HEC-RAS 2-D flood modeling.

Conflicts of Interest: The authors declare no conflict of interest.

References

1. Cannata, M.; Marzocchi, R. Two-dimensional dam break flooding simulation: A GIS-embedded approach. *Nat. Hazards* **2012**, *61*, 1143–1159. [CrossRef]
2. Science Engineering & Sustainability. 2019. Dam Break Simulation with HEC-RAS: Chepete Proposed Dam. Available online: https://sciengsustainability.blogspot.com/2019/03/dam-break-simulation-hec-ras.html (accessed on 5 May 2019).
3. Koskinas, A.; Tegos, A.; Tsira, P.; Dimitriadis, P.; Iliopoulou, T.; Papanicolaou, P.; Koutsoyiannis, D.; Williamson, T. Insights into the Oroville Dam 2017 Spillway Incident. *Geosciences* **2019**, *9*, 37. [CrossRef]
4. Rotta, L.H.; Alcantara, E.; Park, E.; Negri, R.G.; Lin, Y.N.; Bernardo, N.; Mendes, T.S.; Souza Filho, C.R. The 2019 Brumadinho tailings dam collapse: Possible cause and impacts of the worst human and environmental disaster in Brazil. *Int. J. Appl. Earth Obs. Geoinf.* **2020**, *90*, 102119. [CrossRef]

5. Latrubesse, E.M.; Park, E.; Sieh, K.; Dang, T.; Lin, Y.N.; Yun, S.H. Dam failure and a catastrophic flood in the Mekong basin (Bolaven Plateau), southern Laos, 2018. *Geomorphology* **2020**, *362*, 107221. [CrossRef]
6. Dave, P. Hidroituango: Another Landslide Crisis at a Hydroelectric Dam. The Landslide Blog, AGU. 2018. Available online: https://blogs.agu.org/landslideblog/2018/05/21/hidroituango-1/ (accessed on 5 April 2019).
7. Aljazeera News. 2018. Search for Survivors after Deadly Kenya Dam Collapse. Available online: https://www.aljazeera.com/news/2018/05/search-survivors-deadly-kenya-dam-collapse-180511054548148.html (accessed on 5 May 2019).
8. Zin, W.W.; Kawasaki, A.; Takeuchi, W.; San, Z.M.L.T.; Htun, K.Z.; Aye, T.H.; Win, S. Flood hazard assessment of Bago River Basin, Myanmar. *J. Disaster Res.* **2018**, *13*, 14–21. [CrossRef]
9. Mao, J.; Wang, S.; Ni, J.; Xi, C.; Wang, J. Management System for Dam-Break Hazard Mapping in a Complex Basin Environment. *ISPRS Int. J. Geo-Inf.* **2017**, *6*, 162. [CrossRef]
10. Rodrigues, A.S.; Santos, M.A.; Santos, A.D.; Rocha, F. Dam-break flood emergency management system. *Water Resour. Manag.* **2002**, *16*, 489–503. [CrossRef]
11. Tsakiris, G.D. Flood risk assessment: Concepts, modelling, applications. *Nat. Hazards Earth Syst. Sci.* **2014**, *14*, 1361–1369. [CrossRef]
12. Urzică, A.; Mihu-Pintilie, A.; Stoleriu, C.C.; Cîmpianu, C.I.; Huţanu, E.; Pricop, C.I.; Grozavu, A. Using 2D HEC-RAS Modeling and Embankment Dam Break Scenario for Assessing the Flood Control Capacity of a Multi-Reservoir System (NE Romania). *Water* **2020**, *13*, 57. [CrossRef]
13. Albu, L.-M.; Enea, A.; Iosub, M.; Breabăn, I.-G. Dam Breach Size Comparison for Flood Simulations. A HEC-RAS Based, GIS Approach for Drăcșani Lake, Sitna River, Romania. *Water* **2020**, *12*, 1090. [CrossRef]
14. Cho, Y. Application of NEXRAD Radar-Based Quantitative Precipitation Estimations for Hydrologic Simulation Using ArcPy and HEC Software. *Water* **2020**, *12*, 273. [CrossRef]
15. Tahmasbinejad, H.; Feyzolahpour, M.; Mumipour, M.; Zakerhoseini, F. Rainfall-runoff Simulation and Modeling of Karon River Using HEC-RAS and HEC-HMS Models, Izeh District, Iran. *J. Appl. Sci.* **2012**, *12*, 1900–1908. [CrossRef]
16. Horritt, M.; Bates, P. Evaluation of 1D and 2D numerical models for predicting river flood inundation. *J. Hydrol.* **2002**, *268*, 87–99. [CrossRef]
17. Jayakrishnan, R.; Srinivasan, R.; Arnold, J. Comparison of raingage and WSR-88D Stage III precipitation data over the Texas-Gulf basin. *J. Hydrol.* **2004**, *292*, 135–152. [CrossRef]
18. CEIWR-HEC. *HEC-RAS, River Analysis System: Hydraulic Reference Manual*; US Army Corps of Engineers Hydrologic Engineering Center: Davis, CA, USA, 2016.
19. Bhandari, M.; Nyaupane, N.; Mote, S.R.; Kalra, A.; Ahmad, S. 2D Unsteady Flow Routing and Flood Inundation Mapping for Lower Region of Brazos River Watershed. In *World Environmental and Water Resources Congress 2017*; Digital Scholarship@UNLV: Sacramento, CA, USA, 2017; pp. 292–303. [CrossRef]
20. Brunner, M.I.; Seibert, J.; Favre, A. Bivariate return periods and their importance for flood peak and volume estimation. *Wiley Interdiscip. Rev. Water* **2016**, *3*, 819–833. [CrossRef]
21. Yang, Z.; Chen, W.-P. Earthquakes along the East African Rift System: A multiscale, system-wide perspective. *J. Geophys. Res. Solid Earth* **2010**, *115*. [CrossRef]
22. Ayele, A.; Ebinger, C.J.; van Alstyne, C.; Keir, D.; Nixon, C.W.; Belachew, M.; Hammond, J.O.S. Seismicity of the central Afar rift and implications for Tendaho dam hazards. *Geol. Soc. Lond. Spéc. Publ.* **2016**, *420*, 341–354. [CrossRef]
23. Yared, M.G.; Nigussie, T.; Yohannis, B.T. Dam Breach Modeling and Flood Inundation Mapping a Case Study on Kesem Dam. *ACADEMIA*. 2016. Available online: https://www.academia.edu/27150269/Dam_Breach_Modelling_and_Flood_Inundation_Mapping_A_Case_Study_on_Kesem_Dam (accessed on 5 May 2019).
24. CEIWR-HEC; HEC-HMS. *Hydrological Modeling System: Application Guide*; US Army Corps of Engineers Hydrologic Engineering Center: Davis, CA, USA, 2017; pp. 3.1–3.19.
25. MoWIE, Kesem Kebena Dam and Irrigation Project Final Feasibility study and Dam Design Report. Available online: http://www.fao.org/3/ar867e/ar867e.pdf (accessed on 12 April 2019).
26. Ross, C.W.; Prihodko, L.; Anchang, J.; Kumar, S.; Ji, W.J.; Hanan, N.P. HYSOGs250m, global gridded hydrologic soil groups for curve-number-based runoff modeling. *Sci. Data* **2018**, *5*, 180091. [CrossRef] [PubMed]
27. USDA NRCS. *Urban Hydrology for Small Watersheds*; US Dept. of Agriculture, Soil Conservation Service, Engineering Division: Washington, DC, USA, 1986.
28. National Catalog Service for Geographic Information. Globeland30. Available online: http://www.webmap.cn/main.do?method=index (accessed on 3 February 2019).
29. Subyani, A.M. Hydrologic behavior and flood probability for selected arid basins in Makkah area, western Saudi Arabia. *Arab. J. Geosci.* **2011**, *4*, 817–824. [CrossRef]
30. CEIWR-HEC. *Hydrologic Modeling System HEC-HMS, Technical Reference Manual*; U.S. Army Corps of Engineers, Hydrologic Engineering Center (HEC): Davis, CA, USA, 2000.
31. Maidment, D.R.; Morehouse, S. *Arc Hydro: GIS for Water Resources*; ESRI, Inc.: Redlands, CA, USA, 2002.
32. CEIWR-HEC; HEC-GeoHMS. *Geospatial Hydrologic Modeling Extension: Hydraulic Reference Manual*; US Army Corps of Engineers Hydrologic Engineering Center: Davis, CA, USA, 2013.

33. Getahun, Y.S.; Gebre, S.L. Flood hazard assessment and mapping of flood inundation area of the Awash River basin in Ethiopia using GIS and HEC-GeoRAS/HEC-RAS model. *J. Civ. Environ. Eng.* **2015**, *5*, 1. [CrossRef]
34. Behailu, S. Stream Flow Simulation for the Upper Awash Basin. Ph.D. Thesis, Faculty of Technology Department of Civil Engineering, Addis Ababa University, Addis Ababa, Ethiopia, 2004.
35. Kisi, O.; Shiri, J.; Tombul, M. Modeling rainfall-runoff process using soft computing techniques. *Comput. Geosci.* **2013**, *51*, 108–117. [CrossRef]
36. CEIWR-HEC; HEC-GeoRAS. *GIS Tools for Support of HEC-RAS Using ArcGIS. [Online] Version 4.3.93*; US Army Corps of Engineers, Institute of water Resources: Davis, CA, USA, 2011.
37. Froehlich, D.C. Embankment Dam Breach Parameters and Their Uncertainties. *J. Hydraul. Eng.* **2008**, *134*, 1708–1721. [CrossRef]
38. Wahl, T. *Evaluation of Erodibility-Based Embankment Dam Breach Equations (Hydraulic Laboratory Report HL-2014-02)*; US Department of the Interior Bureau of Reclamation: Denver, CO, USA, 2014; p. 99.
39. Thornton, C.I.; Pierce, M.W.; Abt, S.R. Enhanced Predictions for Peak Outflow from Breached Embankment Dams. *J. Hydrol. Eng.* **2011**, *16*, 81–88. [CrossRef]
40. Yu, D.; Lane, S.N. Urban fluvial flood modelling using a two-dimensional diffusion-wave treatment, part 1: Mesh resolution effects. *Hydrol. Process.* **2005**, *20*, 1541–1565. [CrossRef]
41. Chow, V.T. *Open-Channel Hydraulics*; McGraw-Hill Civil Engineering Series; McGraw-Hill Book Company Inc.: New York, NY, USA, 1959.
42. Michaud, J.; Johnson, C.; Iokepa, J.; Marohnic, J. Methods for Estimating the Impact of Hypothetical Dam Break Floods. In *Chemistry for the Protection of the Environment*; Springer: Boston, MA, USA, 2005; pp. 195–199. [CrossRef]
43. Onyutha, C.; Tabari, H.; Taye, M.T.; Nyandwaro, G.N.; Willems, P. Analyses of rainfall trends in the Nile River Basin. *J. Hydro-Environ. Res.* **2016**, *13*, 36–51. [CrossRef]
44. Basheer, T.A.; Wayayok, A.; Yusuf, B.; Kamal, M. Dam Breach parameters and their influence on flood hydrographs for Mosul Dam. *J. Eng. Sci. Technol.* **2017**, *12*, 2896–2908.
45. Awal, R.; Nakagawa, H.; Kawaike, K.; Baba, Y.; Zhang, H. Experimental study on piping failure of natural dam. *J. Jpn. Soc. Civ. Eng. Ser. B1 Hydraul. Eng.* **2011**, *67*, I_157–I_162. [CrossRef]
46. Psomiadis, E.; Tomanis, L.; Kavvadias, A.; Soulis, K.; Charizopoulos, N.; Michas, S. Potential Dam Breach Analysis and Flood Wave Risk Assessment Using HEC-RAS and Remote Sensing Data: A Multicriteria Approach. *Water* **2021**, *13*, 364. [CrossRef]
47. Wu, M.; Ge, W.; Li, Z.; Wu, Z.; Zhang, H.; Li, J.; Pan, Y. Improved Set Pair Analysis and Its Application to Environmental Impact Evaluation of Dam Break. *Water* **2019**, *11*, 821. [CrossRef]

Article

Hydrological and Hydraulic Flood Hazard Modeling in Poorly Gauged Catchments: An Analysis in Northern Italy

Francesca Aureli *, Paolo Mignosa, Federico Prost and Susanna Dazzi

Department of Engineering and Architecture, University of Parma, Parco Area delle Scienze 181/A, 43124 Parma, Italy; paolo.mignosa@unipr.it (P.M.); federico.prost@unipr.it (F.P.); susanna.dazzi@unipr.it (S.D.)
* Correspondence: francesca.aureli@unipr.it

Abstract: Flood hazard is assessed for a watershed with scarce hydrological data in the lower plain of Northern Italy, where the current defense system is inadequate to protect a highly populated urban area located at a river confluence and crossed by numerous bridges. An integrated approach is adopted. Firstly, to overcome the scarcity of data, a regional flood frequency analysis is performed to derive synthetic design hydrographs, with an original approach to obtain the flow reduction curve from recorded water stages. The hydrographs are then imposed as upstream boundary conditions for hydraulic modeling using the fully 2D shallow water model PARFLOOD with the recently proposed inclusion of bridges. High-resolution simulations of the potential flooding in the urban center and surrounding areas are, therefore, performed as a novel extensive application of a truly 2D framework for bridge modeling. Moreover, simulated flooded areas and water levels, with and without bridges, are compared to quantify the interference of the crossing structures and to assess the effectiveness of a structural measure for flood hazard reduction, i.e., bridge adaptation. This work shows how the use of an integrated hydrological–hydraulic approach can be useful for infrastructure design and civil protection purposes in a poorly gauged watershed.

Keywords: flood risk; poorly gauged watersheds; regional flood frequency; flood modeling; GPU-parallel numerical scheme; bridges

Citation: Aureli, F.; Mignosa, P.; Prost, F.; Dazzi, S. Hydrological and Hydraulic Flood Hazard Modeling in Poorly Gauged Catchments: An Analysis in Northern Italy. *Hydrology* **2021**, *8*, 149. https://doi.org/10.3390/hydrology8040149

Academic Editor: Andrea Petroselli

Received: 21 July 2021
Accepted: 30 September 2021
Published: 5 October 2021

1. Introduction

Among the most frequent and destructive disasters, floods annually hit people in many countries all over the world. Flooding is the cause of up to 40% of natural disasters in the world, entailing almost half of the fatalities related to natural hazard, with a rising trend [1–4]. Flood risk management, therefore, has to be implemented on the basis of a proper and aware estimation of flood hazards under the consciousness that, due to global warming, the occurrence of extreme flood events will probably increase in the future, even if some recent studies show a situation that, at least in Europe, seems very patchy [5–8]. This task can be challenging, as it requires the careful consideration of many factors related to catchment properties, hydrological inputs, and characteristics of rural, urban, and productive areas that are potentially floodable [9]. Inaccurate hazard and risk assessments may result in poor risk management interventions, ranging from insufficient protection to squandering of public resources. Accurate flood hazard and risk assessments, on the contrary, can valuably support decisions on land use, design of infrastructure, and drafting and organization of emergency response for civil protection purposes. Inundation mapping allows capturing of the extent of the flooded area and can also represent a tool of primary importance to support first responders [10,11]. Whatever type of model is used for the purpose, a key element in flood risk assessment is the identification of the hydrological stresses to be adopted as input. Unfortunately, the definition of the hydrological inputs is strongly hindered by the fact that, very often, the watersheds of interest are devoid of reliable field observations. Adopting the definition of Sivapalan et al. [12], *ungauged* or *poorly gauged* basins are those in which quality and quantity of hydrological observations are inadequate to allow a reliable evaluation of streamflow. Always the object of ongoing investigation, the prediction or forecasting of the hydrological responses of these basins and the evaluation of the associated uncertainty were the subject of a significant research activity (Prediction in Ungauged Basins—PUB) by the scientific community through the decade 2003–2012 [12]. Rather disheartening results emerged from that analysis, according to which the majority of rivers and stream reaches, and tributaries in the world are ungauged or poorly gauged [13]. Indeed, the knowledge of continuous water stages and discharges in rivers is a crucial factor in watershed analyses and water resource management, due to the need to evaluate flood peaks, flow reduction factors, and duration curves at the basis of the design of engineering structures, such as defense systems, flood detention reservoirs, etc. Very often, unfortunately, even for watersheds with a good consistency of water stage observations, both as regards the length of the measured series and the density of the gauging stations along the river network, the poor reliability or even the unavailability of the stage–discharge relationships affects the potential of the observed database. This could be obviated by developing methodologies capable of extracting useful information for the construction of flow hydrographs directly from water-stage-observed time series.

A possible solution to overcome the lack of direct observations is to resort to regionalization techniques that allow the necessary information to be derived from the knowledge of quantitative data in nearby gauged watersheds or with reference to a region homogeneous with the ungauged or poorly gauged catchment under investigation [14]. Behind the concept of a homogeneous region, there is the assumption that the similitude in climatic, geologic, vegetative, topographic, etc., characteristics would give origin to similar responses in terms of runoff. This does not necessarily imply the neighborhood of the basins [15]. Regionalization can be intended as the process of transferring hydrological information from gauged to poorly gauged watersheds, also, in terms of frequency [16–18]. It is obviously advisable to use any direct information available for the area of interest, even if scarce and fragmentary, to validate, to some extent, its belonging to the identified homogeneous region [19,20]. Many sources of uncertainty can affect the prediction reliability due to data and to the regionalization procedure itself, and this issue is constantly subject to attention given the importance that regionalization procedures play in hydrology [21–24].

In general, once the hydrological inputs of assigned probable frequency have been identified, inundation mapping can be performed through distinct approaches [25]: simplified conceptual procedures, empirical methods, and physically based models. The latter are often run at low resolution to allow reasonable computational times with the aim of creating flooding maps, sometimes also at a continental scale [26–29]. Even if a line of research [30–32] argued that inundation mapping performed at high spatial resolution can lead to useless and even counterproductive detail, there is evidence that high-resolution meshes derived from LiDAR-based DTMs allow a detailed description of the relevant terrain features typical of man-made landscapes (e.g., roads, railways, channels, and embankments) that dominate flow patterns and that would be undetectable at a coarse resolution [33–36]. The use of GPU-accelerated 2D shallow water numerical models and the adoption of nonuniform grids represent a powerful analysis tool, allowing a drastic reduction in the computational costs [11,37]. Care has to be devoted, anyway, to an in-depth analysis of the terrain model, which, even in the presence of high spatial resolution, could be affected by non-negligible descriptive deficiencies, such as those occurring during vegetation filtering along streams with densely vegetated banks [38]. Anyway, the most accurate results for flood modeling are obtained adopting a fine spatial resolution, also, thanks to the accurate description of local terrain features and hydraulic structures that can be achieved [34,39].

Flood propagation in rivers and channels is significantly affected by the presence of bridges and other crossing structures, which represent an obstruction to the flow due to lateral and vertical constrictions and increase the hazard in upstream areas as a consequence of the backwater effect. Moreover, in the presence of high return period flood events, bridge decks may even experience overtopping; it is, therefore, appropriate that all the elements interfering with the flow would be properly considered in numerical simulations aimed at flood management purposes [40]. Different approaches can be adopted to include the presence of bridges in 2D shallow water models [41,42], and the choice of the modeling approach should also consider the possible additional computational burden consequent to the introduction of the necessary solution algorithms. A possible way to simulate the presence of bridges and other hydraulic structures in 2D domains is the implementation of internal boundary conditions [43]. Recent studies investigated the possibility to also implement this approach in GPU-accelerated numerical codes [44], and the validations performed provided good results, both in low- and high-flow conditions for bridges, without affecting the performance of the calculation tools. It is, in fact, very important to preserve the computational efficiency of the numerical codes, both with a view to effectively simulate flooding phenomena on very extensive computing domains and to achieve efficient and accurate 2D simulations in ever shorter calculation times.

With the aim of assessing flood hazard in a poorly gauged watershed in Northern Italy, an integrated hydrologic–hydraulic approach is adopted here. A regional flood frequency analysis is performed to derive synthetic design hydrographs (SDH) to be imposed as upstream boundary conditions for fully 2D high-resolution hydraulic modeling in a domain in which an urban center is located at the confluence of two streams. An original approach is studied to obtain indispensable information for the construction of SDHs by exploiting the knowledge of the behavior of water stage hydrographs instead of discharge ones. Water stage hydrographs are, indeed, usually more accessible and reliable than discharge hydrographs due to the unavailability or uncertainty of the stage–discharge relationships.

The area of interest is characterized by the presence of over 20 crossing structures. This then leads to an extensive application of bridge modeling in a truly 2D framework for each different scenario. Thanks to the high efficiency of the code architecture and bridge computational approach, this entails only a negligible increase in the calculation times compared to the simulation of the same scenario in the absence of the crossing structures. Therefore, the computed flooded areas and water levels, in the actual state and in the hypothetical absence of bridges, are compared for each return period to quantify the influence exerted by the interfering structures upon flooding extent and to assess the effectiveness of a structural measure for flood hazard reduction, i.e., bridge rebuilding.

The paper is organized as follows. The study area is introduced in Section 2, while the hydrological analysis is presented in Section 3. In Section 4, the characteristics and set-up of the model and of the different flooding scenarios considered are illustrated. The results thus obtained are presented in detail in Section 5. The discussion and conclusions are then drawn in Section 6.

2. Study Area

The Chiavenna is a left tributary of the Po River (the main Italian watercourse). It is about 52 km long and flows in the northwest of the Emilia-Romagna region (Northern Italy); its narrow and elongated catchment, which also collects the waters of the Riglio and Chero streams, has a total area of 340 km^2 (Figure 1 and Table 1), about 40% of which is hilly and mountainous. The character of the river system is typically torrential, and not infrequently, in summer, the riverbeds are totally dry.

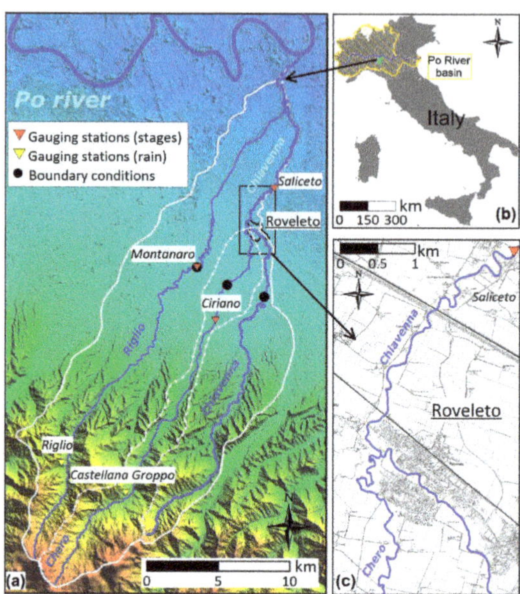

Figure 1. (**a**) Chiavenna watershed, (**b**) location of the Po and Chiavenna basins in Italy, and (**c**) river network in the urban area of Roveleto.

Table 1. Watershed characteristics and mean annual rainfall for contributing areas at the gauging stations of interest for the period 1997–2018.

Watercourse	Gauging Station	Area, A (km^2)	Main Branch Length, L_{mb} (km)	Average Altitude [1], H_{med} (m)	Mean Annual Rainfall (mm)
Chero	Ciriano	56	25	323	962
Chiavenna	Saliceto	161	37	235	876
Riglio	Montanaro	116	30	227	846

[1] Above the gauging station.

3. Hydrological Analysis

With the aim of assessing the hydrological inputs for the hydraulic simulations, design hydrographs of an assigned return period have to be evaluated. They can be completely determined once the peak discharge frequency distribution and the shapes of the hydrographs are identified at the sites of interest. Of course, these features should be inferred from the analysis of historical flood hydrographs at the chosen boundary condition sections, if available. However, the fragmentary or not totally reliable knowledge of the historical flood waves makes it necessary to partly resort to regional techniques.

3.1. Available Data

The historical rainfall data available for the Chiavenna basin refer to the gauging stations of Castellana Groppo (1983–2001) and Riglio (2003–2018). The available records at the two sites were joined into a unique sample of 35 years due to the proximity and similar properties. The gauging sites are, in fact, only about 5 km away (Figure 1) and at the same altitude above sea level (434 m and 432 m for Castellana Groppo and Riglio, respectively), without any noteworthy difference in topographic parameters, such as slope gradient and exposition, capable of significantly influencing the local rainfall patterns. The main characteristics of the rainfall data samples confirm the possibility of merging them in a unique sample (Table 2).

Table 2. Characteristics of the rainfall data samples.

Data Sample	Period	Size	1-h Rainfall		Daily Rainfall	
			Mean (mm)	St. Dev. (mm)	Mean (mm)	St. Dev. (mm)
Castellana Groppo	1983–2001	19	24.3	11.2	66.9	14.5
Riglio	2003–2018	16	23.9	9.5	64.6	17.0
Unique	1983–2018	35	24.1	10.3	65.9	15.5

With reference to the analysis of the historical rainfall unique sample, it is worthwhile stressing two significant values, namely the average values of the 1-h $m(h_1)$ and daily $m(h_d)$ maximum annual rainfall, equal to 24.1 mm and 65.9 mm, respectively (Table 2). These values may be of some interest if regionalization techniques at the outlet of contributing areas along the watercourses have to be adopted in the absence of reliable direct flow records.

Three hydrometric stations are present on the three main streams (Figure 1). For these outlets, watershed main characteristics, together with the mean annual rainfall are reported in Table 1.

For the gauging station on the Chero at Ciriano, frequent lacks and anomalies throughout the available water stage records (since 2002) and the absence of a reliable stage–discharge relationship made the data hardly usable.

The water stage records on the Riglio at Montanaro, instead, provide sufficient accuracy and completeness to determine average values. In particular, a series of 18 historical flood events can be identified. Unfortunately, for this station, the official stage–discharge relationships show excessive variability over the years, probably due to the difficulty in acquiring direct discharge measures. It was, therefore, decided not to rely on the flow records derived through the stage–discharge conversions, but, rather, only on the water stages.

Moreover, the water stages recorded on the Chiavenna at Saliceto appear to be of fairly good quality, and this encourages their analysis with the aim of determining SDHs. As for the previous gauging station, however, the stage–discharge relationships show discordant and sometimes anomalous trends over the years (Figure 2). For this reason, a rating curve was derived through 1D hydraulic modeling, thanks to the availability of river cross-sectional surveys made specifically for this study. Values for Strickler's roughness coefficient equal to 30 and 15 $m^{1/3}$ s^{-1} were assumed for the main channel and the floodplains, respectively, and a monomial function of the form $Q = ah^b$ was fitted to the numerical stage–discharge relation computed at the gauging site (a = 5.579 m^{3-b} s^{-1}, and b = 2.267). From Figure 2, it can be observed that this numerical rating curve is in fairly good agreement with the 2009 official stage–discharge relationship. From the conversion of the recorded hydrographs using the interpolated stage–discharge relationship, the average value of the annual maximum series (AMS) of N flood peaks (N = 18 events) was evaluated, resulting in 94.8 m^3 s^{-1}. It is good to keep this value in mind for considerations that will be made in the following.

Due to the size of the available sample, the observed flood hydrographs can, of course, be trusted for the evaluation of reliable average values and for the identification of a shape of the hydrographs typical of the site of interest. On the other hand, such a reduced sample size is not enough to allow a reliable estimate of maximum values for the high return periods to be investigated in flood hazard assessment. It is, therefore, advisable to resort to alternative methodologies. As a consequence, the peak flow quantiles were estimated through a regional analysis, which requires the evaluation of an index flood and of a growth factor, a dimensionless function increasing with the return period, that allows the identification of the peak flow values for the return periods of interest once multiplied by the index flood.

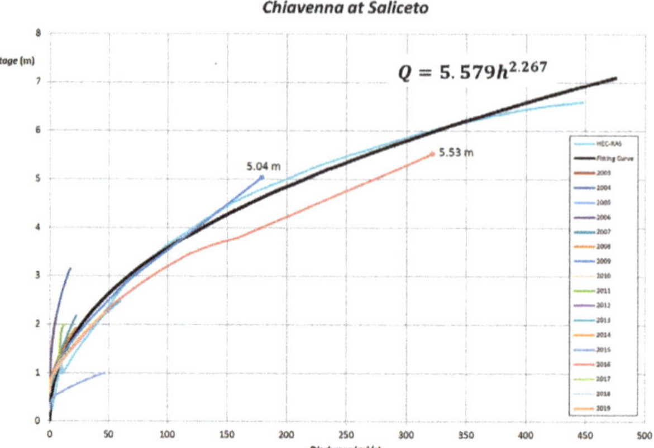

Figure 2. Stage–discharge relationships for the Chiavenna at Saliceto.

3.2. Peak Discharge Estimation

3.2.1. Index Flood

Regionalization procedures allow exploitation of the observations available for a group of watersheds sharing similar behavior and gathered in a homogeneous region to which the catchment of interest is believed to belong, due to geographic location or characteristics. In general, a regionalization procedure is accomplished, first of all, through the determination of a flow scale factor proper of the site of interest, independent of the return period, namely the index flood [45,46]. Index flood is often assumed to be equal to the average of the annual maximum peak flows at the site of interest. Alternatively, it is possible to resort to indirect methods. Among others, multi-regressive models that express the index flood as a function of some significant features of the basin and of the watercourse can be adopted [45–48].

With reference to [45–48], two multi-regressive expressions for the index flood at the river sections of interest were evaluated. In the first, the area A of the catchment, the length L_{mb} of the main branch, and the hourly rain value $m(h_1)$ are taken into consideration:

$$Q_{I1} = 2.797 \cdot 10^{-5} A^{1.235} m(h_1)^{3.513} L_{mb}^{-0.72}, \quad (1)$$

while, in the second, the index flood is calculated with reference to the area A of the catchment, the average altitude H_{med} of the basin at the section of interest, and the pluvial average value relating to the daily rainfall $m(h_d)$:

$$Q_{I2} = 2.1 \cdot 10^{-4} A^{1.082} m(h_d)^{2.416} H_{med}^{-0.469} \quad (2)$$

For the Chiavenna basin (from [45,49]), it is possible to estimate the values of 26 mm and 66 mm, respectively, for $m(h_1)$ and $m(h_d)$. These values are in very good agreement with the corresponding values of 24.1 mm and 65.9 mm obtained from the rain sample introduced in Section 3.1.

The application of Equations (1) and (2) leads to the values shown in Table 3. For the Chiavenna at Saliceto, it is observed that the index flood obtained with the two different multi-regressive relationships is in good agreement with the value obtained analyzing the at site flood waves, equal to 94.8 $m^3\ s^{-1}$, evaluated in Section 3.1. As a precautionary choice, the maximum values provided by the regressive relations at the sections of interest, which correspond to the outcome of Equation (1), will be adopted (bolded in Table 3).

Table 3. Index floods for the gauging stations of interest (selected values in bold type).

Watercourse	Gauging Station	Index Flood Equation (1) Q_{I1} (m^3 s^{-1})	Index Flood Equation (2) Q_{I2} (m^3 s^{-1})
Chero	Ciriano	37	27
Chiavenna	Saliceto	103	98
Riglio	Montanaro	80	70

3.2.2. Growth Factor

The VA.PI. project [48] and the following in-depth studies divided the Italian territory into different macro-regions that were, in turn, subdivided into homogeneous zones, each characterized by a different growth factor. For the different areas of the macro-regions, the growth curves were obtained from the analysis of the multiple gauging stations present in the considered territory and provided with reliable stage–discharge relationships. The growth factor is identified on the basis of a GEV distribution having expression:

$$\chi_T = \xi + \frac{\alpha}{\kappa}\left[1 - e^{-\kappa y}\right] \quad (3)$$

in which y is the Gumbel reduced variate. With regard to Zone C, the closer to the Chiavenna basin, the three parameters ξ, $\alpha$and κ in Equation (3) assume the values 0.643, 0.377, and −0.276, respectively.

Other studies have been based on an updated version of the original database of the VA.PI. project and have further divided the Emilia-Romagna–Marche region into sub-regions, each with its own growth curve [45–47,50].

The growth curve relating to the Zone C region can be compared with that deduced from the analysis of the series of maximum annual peak flows in Saliceto. The sample is well interpreted by a GEV distribution from which a growth factor close to that of Zone C is derived, as shown in Figure 3.

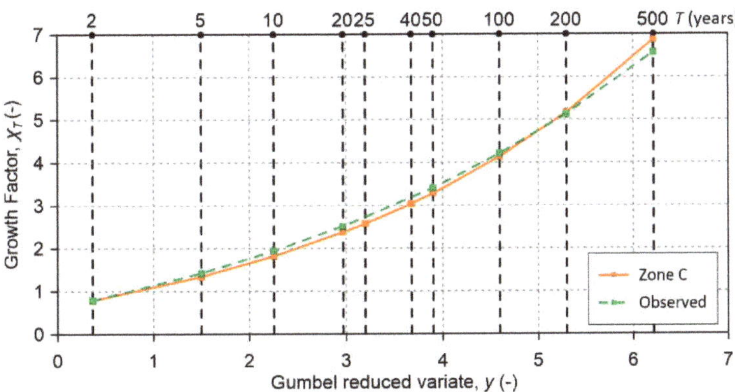

Figure 3. Growth factors as a function of the return period T.

From the index floods and growth factors derived from the previous analyses, the peak discharges were obtained at the sites of interest simply by multiplying the maximum index flood by the growth factor for the chosen return periods of 20, 50, 200, and 500 years (Table 4).

Table 4. Peak discharges for the chosen return periods at the sites of interest.

Watercourse	Gauging Station	Max. Index Flood $Q_{I\,Max}$ (m^3 s^{-1})	Peak Discharges			
			Q_{20} (m^3 s^{-1})	Q_{50} (m^3 s^{-1})	Q_{200} (m^3 s^{-1})	Q_{500} (m^3 s^{-1})
Chero	Ciriano	37	88	122	191	254
Chiavenna	Saliceto	103	245	339	532	707
Riglio	Montanaro	80	190	263	413	549

3.3. *Definition of the Design Hydrographs*

3.3.1. Flow Reduction Curve and Peak Position Ratio

In order to evaluate hazard and risk maps, it is also necessary to estimate the characteristic volumes of the floods and their shape at the site of interest. These features are well represented by the reduction factor $\varepsilon_D(T)$:

$$\varepsilon_D(T) = \frac{\overline{Q}_D(T)}{Q_0(T)} \tag{4}$$

where $Q_0(T)$ is the peak discharge of T years return period and $\overline{Q}_D(T)$ is the average discharge over the duration D of the same return period, and by the position r_D ($0 \leq r_D \leq 1$) of the peak of the hydrographs in the same intervals (Figure 4). Some methodologies for identifying the reduction factor and peak position are available for instrumented river sections, where a series of historical floods has been recorded, and for which a reliable stage–discharge relationship is available. In the absence of the second condition, the reduction curve can be inferred, after appropriate processing, on the basis of the trend of the water stages instead of discharges. If water stages are also not available, as for the majority of non-instrumented river sections, some indirect methods still allow the estimation of the reduction factor and peak position through empirical regional relations, which express them as a function of some characteristics of the basin under consideration [51].

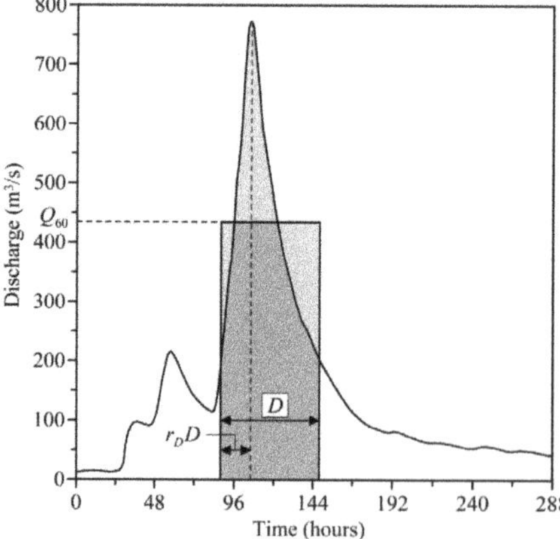

Figure 4. Data sampling of Q_D and r_D from a historical hydrograph (D = 60 h).

In the present case, the water stage series recorded at Saliceto was analyzed directly and after its conversion into discharges through the rating curve previously derived. By analyzing the N available flow hydrographs with N_D moving windows, it was then possible to extract N_D samples with the size N for the value of the maximum annual average flow and ($N_D - 1$) samples for the peak position ratio for the durations of interest (for the duration 0 the position of the peak is, in fact, not defined). The choice of the maximum window duration D_f must derive from a preliminary examination of the characteristic durations of the most significant historical flood hydrographs. For the Chiavenna stream at the gauging site of Saliceto, durations from 0 to 96 h on an hourly basis were considered (D_f = 96 h, N_D = 97).

Although, in the general case, the reduction ratio ε_D is a function of duration D and return period T, in many practical cases, it can be assumed independent of the latter. This is strictly true only if—neglecting the influence of the statistical moments higher than the second—the coefficient of variation $CV(\overline{Q}_D)$ and the probability distribution type of $\overline{Q}_D$ can be considered independent of D. Under these assumptions, ε_D becomes independent of T and reduces to the ratio of the averages of $\overline{Q}_D$ and Q_0:

$$\varepsilon_D = \frac{\mu(\overline{Q}_D)}{\mu(Q_0)} \tag{5}$$

The expression of the reduction curve through a continuous and differentiable function of D, depending on the fewest possible parameters, can be advantageous. Based on the crossing properties of a given threshold value from continuous Gaussian stationary stochastic processes, [52] derived the following theoretical formulation:

$$\varepsilon_D = \sqrt{\frac{\theta}{2D}\left[2 + e^{-\frac{4D}{\theta}} - \frac{3\theta}{4D}\left(1 - e^{-\frac{4D}{\theta}}\right)\right]} \tag{6}$$

Besides the advantages coming from a theoretical basis and from the presence of a unique time parameter θ, Equation (6) is particularly suitable to fit the empirical reduction ratios of large catchments. Moreover, [53] showed that θ can be strictly correlated to the time of concentration of the catchment.

For medium-size catchments (area between 100 and 1000 km^2) [54], like the one considered here, Equation (6) does not accurately fit the empirical reduction factor (Figure 5). It was, therefore, decided to adopt the following generalized form:

$$\varepsilon_D = \left[\frac{\theta}{2D}\left[2 + e^{-\frac{4D}{\theta}} - \frac{3\theta}{4D}\left(1 - e^{-\frac{4D}{\theta}}\right)\right]\right]^{\beta} \tag{7}$$

The higher flexibility given by the two parameters (θ and β) allows a better fit for the sample values (here β = 0.7 and θ = 7.95 h) (Figure 5). It is worth noting that other commonly adopted two-parameter functions, such as the one suggested by [16]:

$$\varepsilon_D = (1 + \gamma D)^{-\delta} \tag{8}$$

where γ and δ are positive parameters, do not fit the empirical data in this case as well (Figure 5).

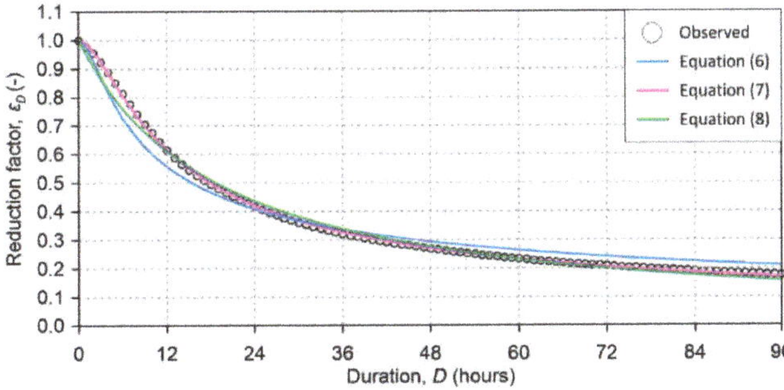

Figure 5. Reduction factor at Saliceto gauging station.

An interesting result can be obtained applying the same procedure to the water stage time series instead of to the discharge time series. If the stage–discharge relationship can be expressed by means of a monomial function of the type adopted in Section 3.1, a very good agreement between the reduction curve determined from the discharges and the one determined on the basis of the flood stages is achieved if the following transformation is adopted:

$$\varepsilon_{D|Q} \cong \varepsilon_{D|h}{}^{\lambda b} \tag{9}$$

where b is the exponent of the stage–discharge relationship expressed in monomial form and λ is a corrective coefficient close to 1. For the Chiavenna at Saliceto, Equation (9) is satisfied in a very good way when λ is equal to the value 0.91 (Figure 6).

This result, even if it must be confirmed with further studies, allowed expansion of the analysis, also, to the station on the Riglio stream, for which a reliable stage–discharge relationship was not available.

In the almost total lack of useful observations for the Chero stream, and given the similarity among the watercourses under investigation, it was decided to adopt the same flow reduction factor and peak position ratio relations (i.e., the same waveform) obtained for the Saliceto gauging station for the Chero stream also.

With regard to the expression of the peak position ratio introduced above, a continuous and differentiable function of D can be useful. N_D time series of observed peak position ratios are available for the Saliceto gauging station, one for each duration considered. For the purpose of the evaluation of the SDH, the average value of each series was calculated and the N_D values thus achieved were interpolated with the expression:

$$r_D(D) = a_1 + \frac{a_2}{a_3 + (D)^{a_4}} \tag{10}$$

imposing $r_D(0) = 0.5$ (Figure 7).

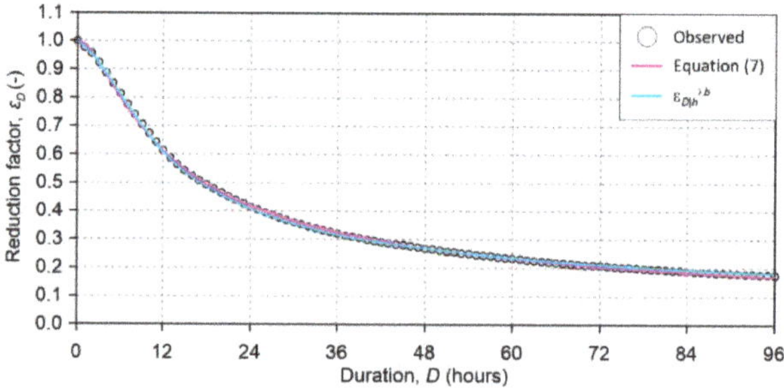

Figure 6. Reduction factor expressed through the stage reduction factor at Saliceto.

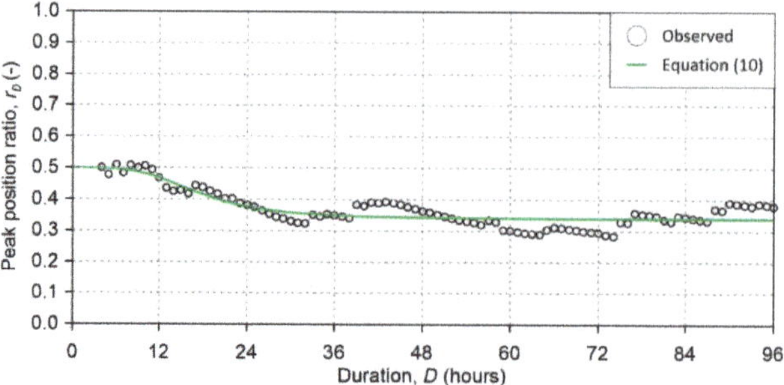

Figure 7. Observed and interpolated peak position ratio.

3.3.2. Evaluation of the SDHs

Following [55], the evaluation of the synthetic design hydrographs (SDHs) is carried out by imposing that the maximum average discharge in each duration coincides with the one predicted by the reduction curve. The shape of the hydrographs is then determined by the peak position ratio behavior.

The synthetic hydrograph is, therefore, defined by the conditions:

$$\int_{-r_D D}^{0} Q(\tau;T)d\tau = r_D\overline{Q}_D(T)D; \quad \int_{0}^{(1-r_D)D} Q(\tau;T)d\tau = r_D(1-r_D)\overline{Q}_D(T)D \tag{11}$$

The evaluation of the two branches of the hydrograph $Q(t; T)$, before and after the peak, is obtained by differentiating Equation (11) with respect to duration D [55] once the expression for the quantiles of the average maximum flow rate as a function of the reduction factor is introduced:

$$\overline{Q}_D(T) = \varepsilon_D Q_0(T) \tag{12}$$

Once the index flood, the growth factor, the reduction factor, and the peak position are evaluated, it is possible to identify the set of SDHs for each section of interest and for the chosen return periods. As an example, Figure 8 shows the SDHs obtained through the adoption of the described procedure for the gauging station of Saliceto.

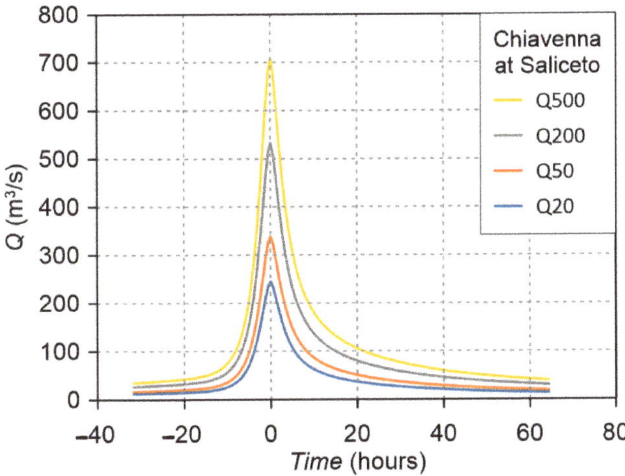

Figure 8. Design hydrographs obtained for the Chiavenna stream at Saliceto gauging station.

3.3.3. Water Stage Record Behavior

Riglio, Chero, and Chiavenna streams are usually subject to simultaneous floods. Negligible time lags between the peaks of the flood hydrographs, in the order of one to three hours, were, in fact, inferred by observing the historical water stage records (Figure 9). In the setting up of flooding scenarios, a synchronous behavior of the hydrological inputs will, therefore, be assumed.

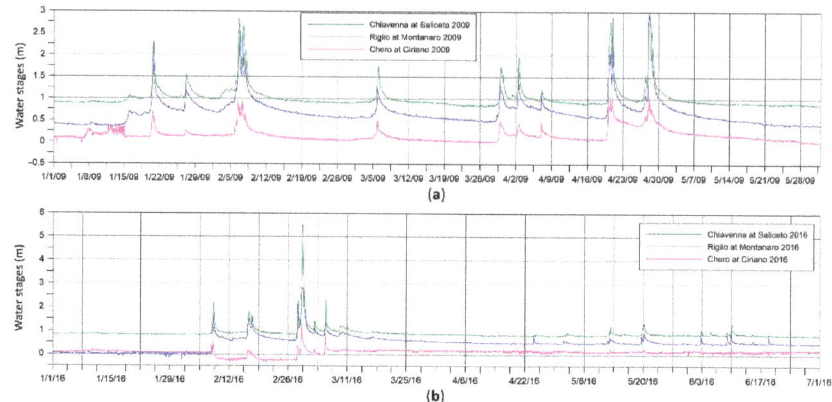

Figure 9. Water stages at the gauging stations of interest for two semesters in (**a**) 2009 and (**b**) 2016.

4. Hydraulic Analysis

Hydraulic analyses focused on the hydrographic network, including the main reach of the Chiavenna stream and of the Chero and Riglio tributaries (Figure 10). The Riglio stream flows in a rather engraved riverbed and is provided with discontinuous embankments of modest height (1–2 m). The Chero stream has almost negligible embankments (<1 m) and runs through an alluvial plain characterized by a rather modest incision, unlike the Chiavenna, which flows upstream of Roveleto in a rather deep valley. The levee system of the Chiavenna stream originates at Roveleto and develops for about 10 km, with increasing height up to the confluence with the main Po River. All watercourses are crossed by several bridges.

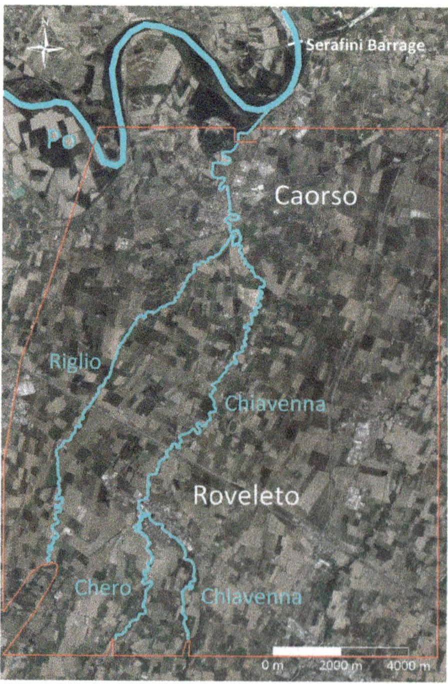

Figure 10. Modeled river network.

4.1. Numerical Model

4.1.1. Hydraulic Model

The 2D hydraulic simulations were performed using the PARFLOOD code [37,39], developed at the University of Parma. This numerical model is based on an explicit finite volume discretization of the fully 2D shallow water equations (SWEs) [56], expressed using the well-balanced formulation proposed by [57]. The model is second-order accurate in both time and space, thanks to the adoption of the second-order Runge–Kutta method and of a depth-positive MUSCL extrapolation, while fluxes are evaluated using the HLLC approximate Riemann solver [56]. The model is compatible with both Cartesian grids and structured nonuniform grids, named Block-Uniform Quadtree (BUQ), as detailed in [37]. The adoption of an unevenly distributed spatial resolution is useful for reducing the number of computational cells in the domain and, consequently, the computational burden, while ensuring high accuracy in the areas of greatest interest (e.g., rivers, channels, buildings, hydraulic structures, embankments). Moreover, the code is developed in the compute unified device architecture (CUDA) environment, which enables parallel computing on graphics processing units (GPUs), leading to a drastic reduction in runtimes (of about two orders of magnitude) compared to serial codes, even for domains of several million cells [39]. The good performances of the PARFLOOD model in both simulations of theoretical cases and practical applications over complex bathymetries are well documented in previous works [11,34,58–62], to which the reader is referred for further details.

4.1.2. Bridge Modeling

Among the approaches available for the inclusion of bridges and other hydraulic structures in the 2D numerical models, internal boundary conditions (IBC) were favored in PARFLOOD thanks not only to their applicability to different flow conditions and structure types, but also to their suitability to predict the field-scale backwater effects induced by the fluid/structure interaction in case of high flows.

The IBC implementation is described in detail by [44] and, here, only very briefly recalled. In summary, each bridge is represented as a segment in the plane, which separates "upstream" and "downstream" IBC cells. The average water levels upstream and downstream of the bridge are computed over these cells, and then used to identify the current flow condition (free-surface flow, partially or fully pressurized flow, overtopping, Figure 11). If levels are below the bridge deck, the IBCs are not activated and only the bridge pier obstructions are considered. Otherwise, the discharge flowing through the structure is evaluated using available formulae from the literature [63] and imposed as specific discharge values at upstream IBC cells (redistributed according to the area available to flow). The same values are imposed in the corresponding downstream IBC cells, while water surface elevations are not modified, so that mass conservation is not impaired. Finally, the special flux treatment adopted at the edges between upstream and downstream IBC cells allows the bridge discontinuity to be naturally taken into account. An example of the application of IBCs to bridges in urban flood modeling is reported in [61].

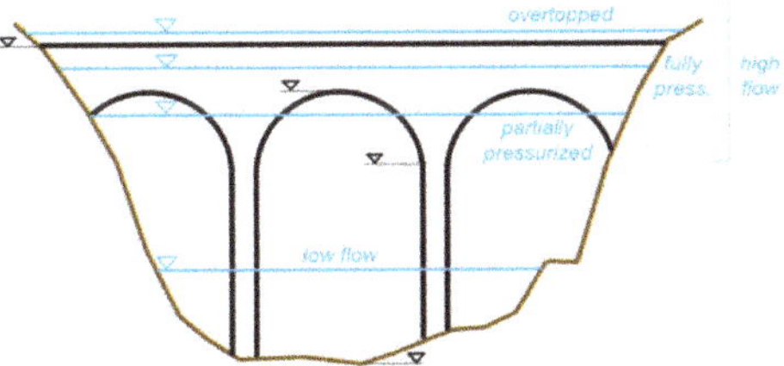

Figure 11. Sketch of a bridge and flow conditions.

4.2. Model Setup

4.2.1. DTM and Building Treatment

The study area covers over 180 km^2. A recent LiDAR survey provided a digital terrain model (DTM) with a resolution of 1 m × 1 m, which was converted to a raster map with a grid size of 2 m × 2 m, considered adequate for this test case. In the grid coarsening process, particular attention was paid to preserving the elevation of retaining structures along the streams and other thin linear topographic features in the domain. A further preprocessing of the DTM was necessary to restore the embankment crest elevations that were not correctly described due to the removal of the bank vegetation cover from the raw LiDAR data, and to integrate in the DTM the bathymetric portion of the riverbed not correctly detected due to the presence of water at the time of the survey.

The presence of buildings in urban areas was included by adopting the building hole (BH) approach [64]. This strategy requires that the computational cells falling within the footprints of buildings are removed from the mesh, which may be achieved by superimposing the shapefile containing the outlines of buildings on the DTM. Such a detailed building treatment is possible thanks to the high mesh resolution. Moreover, a total number of 21 bridges (Figure 12) were introduced in the hydraulic model.

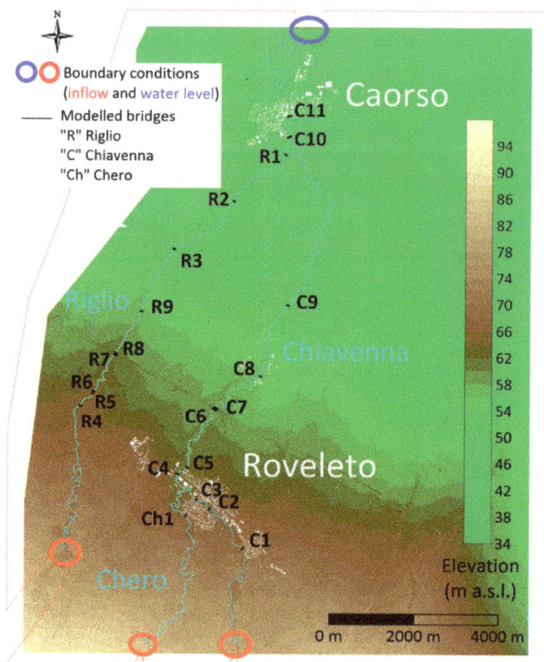

Figure 12. DTM of the hydraulic model and position of boundary conditions and modeled bridges.

4.2.2. Mesh

Starting from the DTM at 2 m × 2 m, a multiresolution grid with cells of variable size, from a minimum value of 2 m to the maximum of 2^4 = 16 m, was built. In order to model the flood propagation accurately, all the main waterways, urban areas, and road and rail communication routes were described with the highest resolution (Figure 13). Elsewhere, the spatial resolution is automatically relaxed by the preprocessing algorithm, as described in [37]. The calculation grid thus identified is made up of a total of 6.8 million cells.

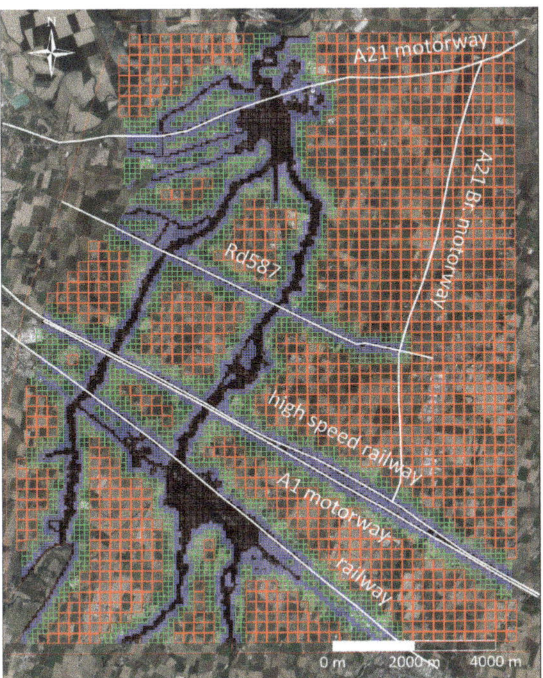

Figure 13. Multiresolution grid and main infrastructures present in the modeled domain; 2 m × 2 m cells in black, 4 m × 4 m cells in blue, 8 m × 8 m cells in green, and 16 m × 16 m cells in red.

4.2.3. Domain Roughness

For the river beds, in the absence of data for calibration, a value of the Strickler's roughness coefficient of 25 $m^{1/3}$ s^{-1} was assumed based on local inspections, literature suggestions [65], and previous studies conducted by the authors concerning neighboring watersheds with similar characteristics [62]. Moreover, this value allowed the reproduction of, at best, the numerical rating curve previously obtained at Saliceto through 1D hydraulic modeling.

For the flooded areas, data for the calibration of roughness are usually unavailable. In literature applications, values for rural areas are highly variable, covering a range between 10 and 40 $m^{1/3}$ s^{-1} [66–69]. For the low plain areas of Emilia-Romagna, indications can be obtained from the work by [34], where a Strickler's coefficient equal to 20 $m^{1/3}$ s^{-1} was calibrated to reproduce the well-known dynamics of the inundation caused by an embankment breach. Preliminary simulations of the present study indicated no significant differences in maximum water depths if values equal to 20 or 25 $m^{1/3}$ s^{-1} were adopted outside the riverbeds, while flooding propagation times are slightly more influenced. As far as the time of arrival of the flooding is important in the emergency management phase during a real event, it is completely irrelevant for the identification of potentially floodable areas. In the absence of further information, it was, therefore, decided to adopt a uniform Strickler's coefficient equal to 25 $m^{1/3}$ s^{-1} over the whole domain.

4.2.4. Boundary and Initial Conditions

The SDHs obtained through the hydrological analysis were imposed as upstream boundary conditions at the sections indicated in Figure 12. At the confluence of the Chiavenna with the Po River, a constant water level was imposed as a downstream boundary condition, due to the presence of a power plant on the Po River just two kilometers downstream of the confluence. The water level upstream of the barrage of the power plant is constantly kept at 41.00 m a.s.l. [70], except for very high floods on the Po River, which never occur simultaneously with the floods of the Chiavenna stream, due to the enormous difference in size of the two river basins.

A steady flow of 10, 15, and 20 $m^3 s^{-1}$ for the Chero, Chiavenna, and Riglio streams, respectively, was assumed as the initial condition.

5. Results

Four flooding scenarios were considered, adopting as inflow boundary conditions the synthetic design hydrographs estimated for T = 20, 50, 200, and 500 years. For the sake of space, only the main results of the study related to the T = 200 years reference scenario (the current flood protection standard in Italy) are presented here and discussed in detail. For this reference scenario, an additional corresponding simulation in the absence of all bridges was performed, in order to evaluate the obstruction effect exerted by these structures. The comparison between the numerical results in the presence and absence of bridges allows an immediate assessment of their influence on the flooding dynamics and on the extent of the flooding upstream and downstream of the urban area of Roveleto.

For the 200 years scenario in the current state, the extension of the flooded areas is remarkable (Figure 14a). Flooding due to the insufficient conveyance of the Chero extends downstream, reaching Roveleto and partly reflowing into the Chiavenna stream (A in Figure 14a). The flooding involves almost the entire town of Roveleto, where pressurized flow is observed at all the bridges, some of which are also overtopped, albeit to a modest extent. The flooded waters also lean against the railway embankment, which is largely overtopped. Proceeding northwards, the flooding extends further downstream, reaching the motorway and the high-speed railway (B in Figure 14a), with depths higher than 2 m. Both the infrastructures are overtopped, and the flooding spreads in the northeast direction towards San Pietro in Cerro and is retained by other motorway embankments. Additional inundations due to outflows from the Riglio stream are also observed, and the water depths are particularly high in the rural area bounded by the two streams, upstream of Caorso. Overall, the scenario is extremely severe in terms of flooded urban and suburban areas and of the crucial transport infrastructures involved.

The same hydrological scenario has been simulated in the hypothetical situation of the absence of all bridges. In this case, a minor extension of the flooding and lower water depths occurs in the areas upstream of the removed crossing structures. At the same time, larger flooding areas are estimated downstream due to the higher discharge released, which exceeds, to a greater extent, the conveyance of the downstream river branches (Figure 14b).

The differences in the flooding extent are even more evident if we focus on the urban area. Figure 15a shows the detail inside Roveleto in the current state, showing that urban areas are totally flooded due to the backwater effect induced by the presence of crossing structures with insufficient conveyance. With reference to the northernmost portion of the inhabited area between the two streams, close to the confluence, the water depths on the ground are still quite low (less than 0.6 m, waterways and underpasses excluded). However, it should be noted that the presence of even a few tens of centimeters on the ground necessarily implies the total flooding of the underground floors, a circumstance that is not immediately evident due to the lack of description of these elements in the DTM.

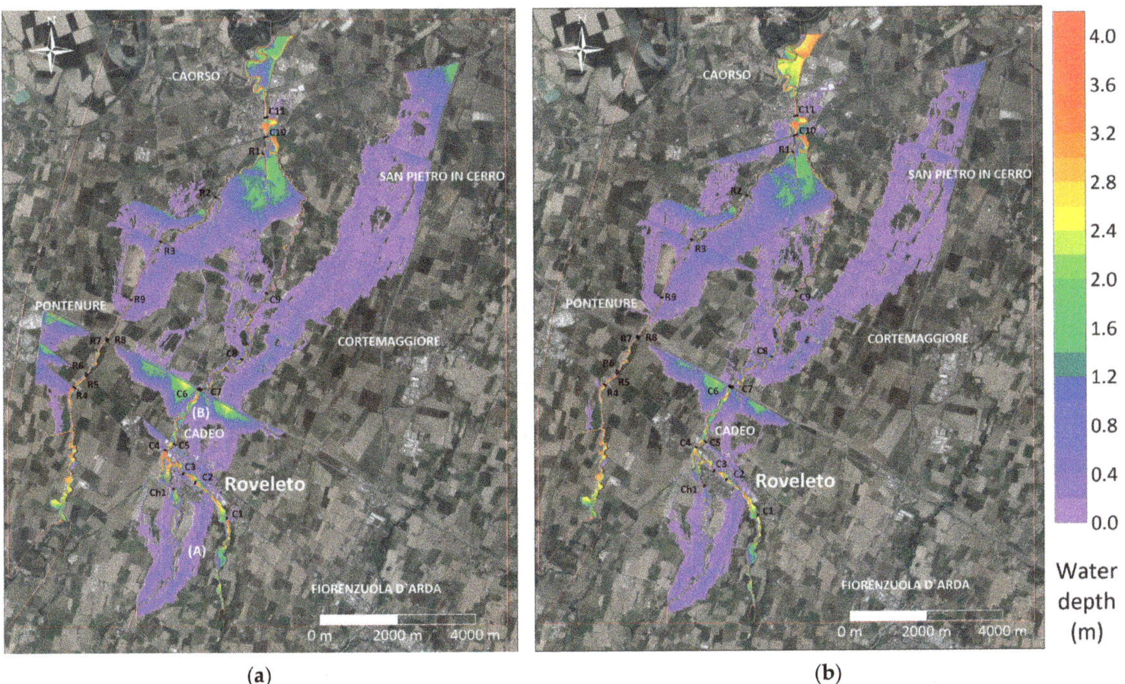

Figure 14. Maximum water depths at Roveleto for $T = 200$ years: (**a**) current condition; (**b**) hypothetical scenario of absence of bridges.

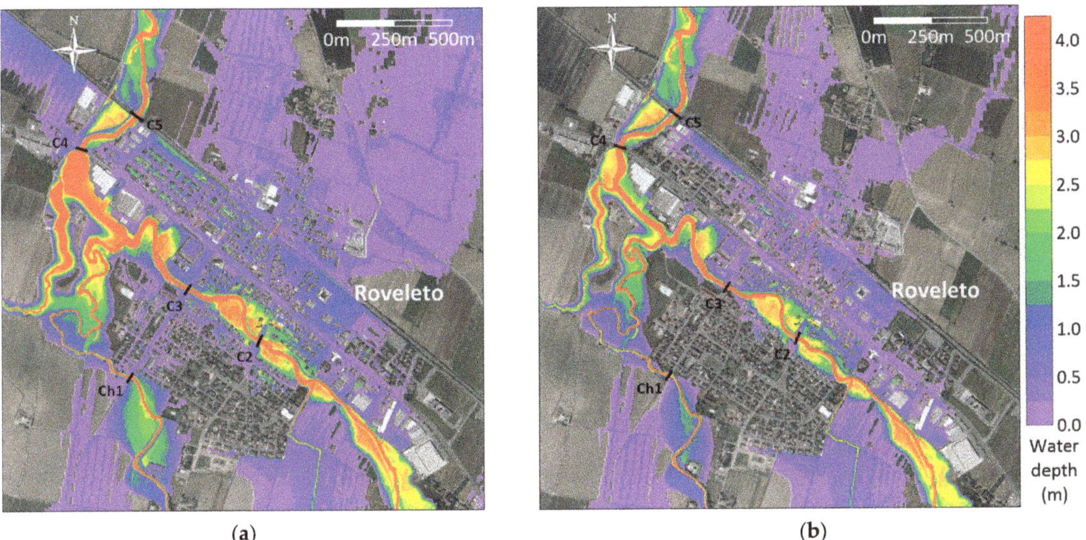

Figure 15. Maximum water depths at Roveleto for $T = 200$ years: (**a**) current condition; (**b**) hypothetical scenario of absence of bridges.

Even in the hypothetical scenario of absence of all the bridges, the majority of the urban area of Roveleto is flooded, but with almost halved water depths compared to the current situation. The reduction in the maximum levels, of the order of several tens of centimeters, would result in a significant reduction in the areas involved, preserving more than half of the urban area from flooding. From Figure 15b, it can be appreciated that the portion of Roveleto, bounded by the Chero and the Chiavenna immediately upstream of their confluence, would be almost entirely unaffected by the flooding. It then follows that, even if the railway is still overtopped, the flooding spreading towards the northeast involves a smaller area. This is obviously due to the lower level of the flooding originating upstream of the highway as a result of the free outflow at the section of the bridge (here removed). The overtopping of the embankments of the highway and of the high-speed railway would be, in this case, minimal. In the absence of bridges, the decrease in maximum water levels ranges from a few tens of centimeters up to one meter and more, with an extreme benefit in terms of population and assets involved. However, the solution of rebuilding the majority of the bridges, increasing their deck level and removing abutments (situation similar to this hypothetical scenario), does not appear to be easily achievable due to the elevation constraints induced by the existing main infrastructures, such as roads, motorways, and railways. Moreover, the removal of the crossing structures, due to the consequent absence of the obstruction effect, allows greater discharges to propagate downstream, giving rise to an increase in water levels in the lower stretch of the stream compared to the current situation, possibly inducing additional outflows. For this reason, the rebuilding of the bridges, besides being difficult to implement, does not even seem appropriate to secure the entire territory from the risk of flooding.

Bridges' Conveyance

Figure 16 shows a few examples of the maximum water levels upstream of six of the more than twenty bridges present in the computational domain (see Figures 12 and 14 for their location).

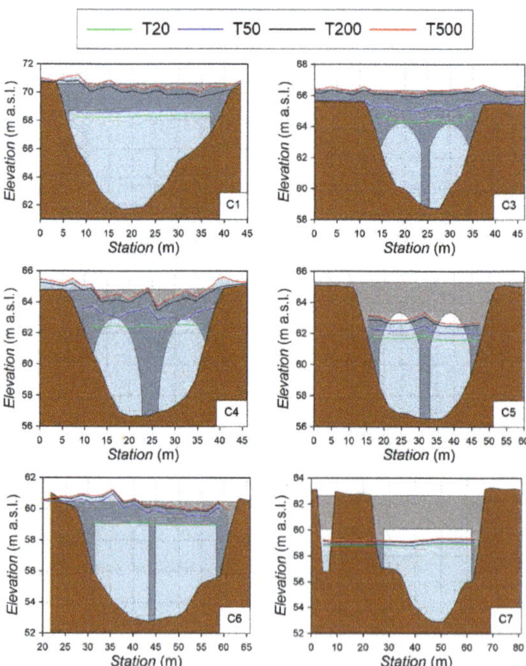

Figure 16. Maximum water levels for some of the crossing structures of interest.

It is worth noting that each profile does not refer to a particular instant, but represents the envelope of all the results obtained during each transient simulation, possibly including flooding outside of the river (see bridge C4 for the 200- and 500-years scenarios). For the most upstream bridges, incipient (C1) or weak (C3) pressurization is already estimated for the $T = 20$ years scenario. The conditions become slightly less severe further downstream (C4) due to the reduced discharge caused by the upstream flooding. For $T = 50$ years, complete pressurization conditions are estimated at the bridges inside Roveleto (C1, C3, and C4). Free outflow instead occurs from the railway crossing onwards (C5, C7), with few exceptions (C6). For the scenario with return period $T = 200$ years, for all bridges inside Roveleto, pressurization and incipient overflow conditions occur. Again, free outflow is observed along the Chiavenna from the railway onwards, with the exception of the motorway bridge that is slightly overtopped on the left side (C6). With reference to the catastrophic scenario for $T = 500$ years, a general worsening of the outflow conditions inside Roveleto is observed. Free outflow is still observed along the Chiavenna at the railway crossing, while the outflow conditions at the bridges further downstream worsen appreciably.

6. Discussion and Conclusions

The case study analyzed here can be considered a reference example of the many challenges faced in hydrologic/hydraulic flood hazard modeling when dealing with poorly gauged watersheds. In this work, some improvements in already consolidated literature methodologies are introduced as regards the hydrologic analysis, while a recently proposed approach for bridge inclusion in fully 2D hydrodynamic models solving the complete SWEs is applied for the first time, with the purpose of assessing the adequacy of bridge conveyance in a complex urban environment.

As regards the hydrologic analysis, due to the scarcity of reliable field observations, the flooding scenarios simulated for different return periods were conducted with reference to synthetic design hydrographs, the peak discharges of which were estimated through a regional approach. Of course, this procedure can lead to results affected by an uncertainty that is difficult to quantify. Nevertheless, the choice to rely on estimates based on regional procedures seems appropriate and promising. Recent studies have, in fact, shown that, in this region, the methodologies adopted for the estimation of the discharge quantiles are characterized by a good reliability [47]. It will be very important in the future to deepen the analyses at a regional scale by expanding the database of direct observations of the reference hydrological variables as much as possible. Some difficulties arising in the fitting of the empirical reduction ratios through the well-known Equation (6) suggested adopting a generalized form of the same relation, in which the presence of an additional parameter allowed a better fit for the available observed values. Moreover, the possibility of deriving the reduction factor on the basis of the analysis of water stages (therefore, not relying on the stage–discharge conversion) could dramatically increase the set of observations available for regional analyses. This would make it possible to exploit historical information available at gauging stations devoid of reliable rating curves, but sometimes rich in several decades of reliable hydrometric water stage records instead. Large-scale analyses that cannot be undertaken at the present time since, in the vast majority of watersheds, only a small percentage of gauging stations is equipped with a reliable stage–discharge relationship would, therefore, be allowed. Both of these aspects seem to be worthy of further investigation.

As regards hydrodynamic modeling, all flooding scenarios are simulated here adopting a GPU-parallelized model based on an explicit shock-capturing finite volume method for the solution of the fully 2D shallow water equations. The river and the flood-prone areas belong to a unique computational domain, and the flow is modeled avoiding the special treatments required by other simplified numerical schemes (quasi 2D, 1D–2D, etc. [71]). Moreover, the hydraulic modeling of the many bridges present in the flow field is here conducted following a truly 2D approach. This prevents significant flow field distortions that can occur in case contiguous crossing structures are modeled following too simplified approaches. Thanks to an efficient parallelization, the issues related to long computational times are overcome, even in the case of high-resolution grids and even when a very large number of crossing structures are introduced. This is of great importance since preserving computational efficiency without affecting the accuracy of the description of the interfering elements is a fundamental objective to lay the foundations for an increasingly advanced ultra-fast modeling that allows real-time detailed simulation to be achieved in the near future.

The strategy of comparing flooding scenarios with and without bridges is used here to investigate possible structural measures to reduce the flood hazard in the study area, e.g., the rebuilding of bridges with increased deck elevation and in the absence of abutments. Indeed, it is evident that most of the bridges are inadequate to convey floods with large return periods. In the absence of bridges, the reduced backwater upstream of the crossing structures would result in a significant decrease in the inundated areas, preserving from flooding at least half of the inhabited center of Roveleto. However, this solution does not appear to be easily practicable due to the elevation constraints induced by the existing infrastructures, roads, and railways. Furthermore, this intervention would not be sufficient to prevent the urban inundation completely, and does not even appear entirely desirable, given the proven worsening of the downstream conditions due to the greater discharge released. Therefore, further structural interventions have to be designed, and the high-resolution 2D model setup for this study can be an extremely powerful tool for assessing the effectiveness of any adaptation measure to reduce the exposure to flood of Roveleto without increasing flood hazard in downstream territories. Examples may include the local adjustments of the embankments' elevations or the identification of temporarily floodable areas upstream of the urban center.

In summary, the flood hazard assessment conducted here with reference to a poorly instrumented watershed has highlighted how the use of an integrated approach of hydrological and hydraulic methodologies can lead to results useful for the design of infrastructures and civil protection purposes. The methodology proposes a series of good practices that can also be applied in those circumstances in which the essential assessment of the flood hazard in highly urbanized areas may, at a first glance, appear strongly discouraged by the scarcity of reliable local hydrological information.

Author Contributions: Conceptualization and methodology, F.A., P.M. and F.P.; software, F.P., F.A. and S.D.; validation, F.P. and F.A.; investigation, F.A. and F.P.; writing—original draft preparation, F.A.; writing—review and editing, F.A., P.M., F.P. and S.D.; visualization, F.A. and F.P.; supervision, F.A., S.D. and P.M. All authors have read and agreed to the published version of the manuscript.

Funding: This research received no external funding.

Institutional Review Board Statement: Not applicable.

Informed Consent Statement: Not applicable.

Data Availability Statement: The data presented in this study are available on request from the first author. The data are not publicly available due to restrictions from the raw data providers.

Acknowledgments: The authors would like to thank Regional Agency for Territorial Security and Civil Protection of Emilia-Romagna, Italy, for supporting this study. This research benefits from the HPC (High-Performance Computing) facility of the University of Parma.

Conflicts of Interest: The authors declare no conflict of interest.

References

1. Noji, E.K. Natural Disasters. *Crit. Care Clin.* **1991**, *7*, 271–292. [CrossRef] [PubMed]
2. Munich Reinsurance. *NatCatSERVICE Loss Events Worldwide 1980–2015*; Munich Reinsurance: Munich, Germany, 2016.
3. Ohl, C.A. Flooding and human health. *BMJ* **2000**, *321*, 1167–1168. [CrossRef] [PubMed]
4. Paprotny, D.; Sebastian, A.; Morales-Nápoles, O.; Jonkman, S.N. Trends in flood losses in Europe over the past 150 years. *Nat. Commun.* **2018**, *9*, 1985. [CrossRef] [PubMed]

5. Alfieri, L.; Feyen, L.; Dottori, F.; Bianchi, A. Ensemble flood risk assessment in Europe under high end climate scenarios. *Glob. Environ. Chang.* **2015**, *35*, 199–212. [CrossRef]
6. Dottori, F.; Szewczyk, W.; Ciscar, J.-C.; Zhao, F.; Alfieri, L.; Hirabayashi, Y.; Bianchi, A.; Mongelli, I.; Frieler, K.; Betts, R.A.; et al. Increased human and economic losses from river flooding with anthropogenic warming. *Nat. Clim. Chang.* **2018**, *8*, 781–786. [CrossRef]
7. Molinari, D.; Dazzi, S.; Gattai, E.; Minucci, G.; Pesaro, G.; Radice, A.; Vacondio, R. Cost–benefit analysis of flood mitigation measures: A case study employing high-performance hydraulic and damage modelling. *Nat. Hazards* **2021**, *108*, 3061–3084. [CrossRef]
8. Blöschl, G.; Hall, J.; Viglione, A.; Perdigão, R.A.P.P.; Parajka, J.; Merz, B.; Lun, D.; Arheimer, B.; Aronica, G.T.; Bilibashi, A.; et al. Changing climate both increases and decreases European river floods. *Nature* **2019**, *573*, 108–111. [CrossRef]
9. Li, W.; Lin, K.; Zhao, T.; Lan, T.; Chen, X.; Du, H.; Chen, H. Risk assessment and sensitivity analysis of flash floods in ungauged basins using coupled hydrologic and hydrodynamic models. *J. Hydrol.* **2019**, *572*, 108–120. [CrossRef]
10. Apel, H.; Aronica, G.T.; Kreibich, H.; Thieken, A.H. Flood risk analyses—How detailed do we need to be? *Nat. Hazards* **2009**, *49*, 79–98. [CrossRef]
11. Ferrari, A.; Dazzi, S.; Vacondio, R.; Mignosa, P. Enhancing the resilience to flooding induced by levee breaches in lowland areas: A methodology based on numerical modelling. *Nat. Hazards Earth Syst. Sci.* **2020**, *20*, 59–72. [CrossRef]
12. Hrachowitz, M.; Savenije, H.H.G.; Blöschl, G.; McDonnell, J.J.; Sivapalan, M.; Pomeroy, J.W.; Arheimer, B.; Blume, T.; Clark, M.P.; Ehret, U.; et al. A decade of Predictions in Ungauged Basins (PUB)—A review. *Hydrol. Sci. J.* **2013**, *58*, 1198–1255. [CrossRef]
13. Razavi, T.; Coulibaly, P. Streamflow Prediction in Ungauged Basins: Review of Regionalization Methods. *J. Hydrol. Eng.* **2013**, *18*, 958–975. [CrossRef]
14. Blöschl, G.; Sivapalan, M. Scale issues in hydrological modelling: A review. *Hydrol. Process.* **1995**, *9*, 251–290. [CrossRef]
15. Smakhtin, V.U. Low flow hydrology: A review. *J. Hydrol.* **2001**, *240*, 147–186. [CrossRef]
16. NERC. *Flood Studies Report*; HMSO London: Swindon, UK, 1975.
17. Michele, C.; De Rosso, R. Uncertainty Assessment of Regionalized Flood Frequency Estimates. *J. Hydrol. Eng.* **2001**, *6*, 453–459. [CrossRef]
18. Bocchiola, D.; De Michele, C.; Rosso, R. Review of recent advances in index flood estimation. *Hydrol. Earth Syst. Sci.* **2003**, *7*, 283–296. [CrossRef]
19. Castellarin, A.; Vogel, R.M.; Matalas, N.C. Multivariate probabilistic regional envelopes of extreme floods. *J. Hydrol.* **2007**, *336*, 376–390. [CrossRef]
20. Castellarin, A.; Burn, D.H.; Brath, A. Assessing the effectiveness of hydrological similarity measures for flood frequency analysis. *J. Hydrol.* **2001**, *241*, 270–285. [CrossRef]
21. Westerberg, I.K.; Wagener, T.; Coxon, G.; McMillan, H.K.; Castellarin, A.; Montanari, A.; Freer, J. Uncertainty in hydrological signatures for gauged and ungauged catchments. *Water Resour. Res.* **2016**, *52*, 1847–1865. [CrossRef]
22. Wagener, T.; Montanari, A. Convergence of approaches toward reducing uncertainty in predictions in ungauged basins. *Water Resour. Res.* **2011**, *47*, 6301. [CrossRef]
23. Zhang, Z.; Wagener, T.; Reed, P.; Bhushan, R. Reducing uncertainty in predictions in ungauged basins by combining hydrologic indices regionalization and multiobjective optimization. *Water Resour. Res.* **2008**, *44*, 1–13. [CrossRef]
24. Hosking, J.R.M.; Wallis, J.R. *Regional Frequency Analysis*; Cambridge University Press: Cambridge, UK, 1997. [CrossRef]
25. Teng, J.; Jakeman, A.J.; Vaze, J.; Croke, B.F.W.; Dutta, D.; Kim, S. Flood inundation modelling: A review of methods, recent advances and uncertainty analysis. *Environ. Model. Softw.* **2017**, *90*, 201–216. [CrossRef]
26. Dottori, F.; Salamon, P.; Bianchi, A.; Alfieri, L.; Hirpa, F.A.; Feyen, L. Development and evaluation of a framework for global flood hazard mapping. *Adv. Water Resour.* **2016**, *94*, 87–102. [CrossRef]
27. Sampson, C.C.; Smith, A.M.; Bates, P.D.; Neal, J.C.; Alfieri, L.; Freer, J.E. A high-resolution global flood hazard model. *Water Resour. Res.* **2015**, *51*, 7358–7381. [CrossRef]
28. Schumann, G.J.P.; Neal, J.C.; Voisin, N.; Andreadis, K.M.; Pappenberger, F.; Phanthuwongpakdee, N.; Hall, A.C.; Bates, P.D. A first large-scale flood inundation forecasting model. *Water Resour. Res.* **2013**, *49*, 6248–6257. [CrossRef]
29. Wing, O.E.J.; Bates, P.D.; Sampson, C.C.; Smith, A.M.; Johnson, K.A.; Erickson, T.A. Validation of a 30 m resolution flood hazard model of the conterminous United States. *Water Resour. Res.* **2017**, *53*, 7968–7986. [CrossRef]
30. Dottori, F.; Di Baldassarre, G.; Todini, E. Detailed data is welcome, but with a pinch of salt: Accuracy, precision, and uncertainty in flood inundation modeling. *Water Resour. Res.* **2013**, *49*, 6079–6085. [CrossRef]
31. Savage, J.T.S.; Pianosi, F.; Bates, P.; Freer, J.; Wagener, T. Quantifying the importance of spatial resolution and other factors through global sensitivity analysis of a flood inundation model. *Water Resour. Res.* **2016**, *52*, 9146–9163. [CrossRef]
32. Savage, J.T.S.; Bates, P.; Freer, J.; Neal, J.; Aronica, G. When does spatial resolution become spurious in probabilistic flood inundation predictions? *Hydrol. Process.* **2016**, *30*, 2014–2032. [CrossRef]
33. Passalacqua, P.; Belmont, P.; Staley, D.M.; Simley, J.D.; Arrowsmith, J.R.; Bode, C.A.; Crosby, C.; DeLong, S.B.; Glenn, N.F.; Kelly, S.A.; et al. Analyzing high resolution topography for advancing the understanding of mass and energy transfer through landscapes: A review. *Earth Sci. Rev.* **2015**, *148*, 174–193. [CrossRef]
34. Vacondio, R.; Aureli, F.; Ferrari, A.; Mignosa, P.; Dal Palù, A. Simulation of the January 2014 flood on the Secchia River using a fast and high-resolution 2D parallel shallow-water numerical scheme. *Nat. Hazards* **2016**, *80*, 103–125. [CrossRef]

35. Bates, P.D. Integrating remote sensing data with flood inundation models: How far have we got? *Hydrol. Process.* **2012**, *26*, 2515–2521. [CrossRef]
36. Zheng, X.; Maidment, D.R.; Tarboton, D.G.; Liu, Y.Y.; Passalacqua, P. GeoFlood: Large-Scale Flood Inundation Mapping Based on High-Resolution Terrain Analysis. *Water Resour. Res.* **2018**, *54*, 10013–10033. [CrossRef]
37. Vacondio, R.; Dal Palù, A.; Ferrari, A.; Mignosa, P.; Aureli, F.; Dazzi, S. A non-uniform efficient grid type for GPU-parallel Shallow Water Equations models. *Environ. Model. Softw.* **2017**, *88*, 119–137. [CrossRef]
38. Anders, N.; Valente, J.; Masselink, R.; Keesstra, S. Comparing Filtering Techniques for Removing Vegetation from UAV-Based Photogrammetric Point Clouds. *Drones* **2019**, *3*, 61. [CrossRef]
39. Vacondio, R.; Dal Palù, A.; Mignosa, P. GPU-enhanced Finite Volume Shallow Water solver for fast flood simulations. *Environ. Model. Softw.* **2014**, *57*, 60–75. [CrossRef]
40. Costabile, P.; Macchione, F.; Natale, L.; Petaccia, G. Comparison of scenarios with and without bridges and analysis of backwater effect in 1-D and 2-D river flood modeling. *Comput. Model. Eng. Sci.* **2015**, *109–110*, 81–103. [CrossRef]
41. Ratia, H.; Murillo, J.; García-Navarro, P. Numerical modelling of bridges in 2D shallow water flow simulations. *Int. J. Numer. Methods Fluids* **2014**, *75*, 250–272. [CrossRef]
42. Maranzoni, A.; Dazzi, S.; Aureli, F.; Mignosa, P. Extension and application of the Preissmann slot model to 2D transient mixed flows. *Adv. Water Resour.* **2015**, *82*, 70–82. [CrossRef]
43. Morales-Hernández, M.; Murillo, J.; García-Navarro, P. The formulation of internal boundary conditions in unsteady 2-D shallow water flows: Application to flood regulation. *Water Resour. Res.* **2013**, *49*, 471–487. [CrossRef]
44. Dazzi, S.; Vacondio, R.; Mignosa, P. Internal boundary conditions for a GPU-accelerated 2D shallow water model: Implementation and applications. *Adv. Water Resour.* **2020**, *137*, 103525. [CrossRef]
45. Brath, A.; Castellarin, A.; Franchini, M.; Galeati, G. La stima della portata indice mediante metodi indiretti. *L'Acqua* **1999**, *3*, 9–16.
46. Brath, A.; Castellari, A.; Franchini, M.; Galeati, G. Estimating the index flood using indirect methods. *Hydrol. Sci. J.* **2001**, *46*, 399–418. [CrossRef]
47. Benassi, P. Stima Della Portata Di Progetto in Bacini Appenninici Non Strumentati: Approcci Regionali a Confronto. Master's Thesis, University of Bologna, Bologna, Italy, 2016.
48. GNDCI. *Progetto Speciale VAPI, Compartimento Bologna-Ancona-Pisa*; Technical Report; CNR-IRPI, GNDCI: Milano, Italy, 2001.
49. DIA, Università di Parma; Servizio Area Affluenti Po; Agenzia Regionale per la Sicurezza Territoriale e la Protezione Civile. *Studio Idrologico Idraulico Finalizzato a Valutare Le Condizioni Di Rischio Delle Aste Sottoposte Al Servizio Di Piena e per Il Nodo Idraulico Di Roveleto Di Cadeo, Analisi Idrologiche*; Technical Report; RER Emilia-Romagna Region: Bologna, Italy, 2019. Available online: https://protezionecivile.regione.emilia-romagna.it/bandi-e-gare/inviti-per-affidamenti/2016/studio_idrolog/studio_idrolog (accessed on 7 July 2021).
50. De Michele, C.; Rosso, R. *Rapporto Sulla Valutazione Delle Piene Italia Nord Occidentale. Portata Al Colmo Di Piena Bacino Del Fiume Po e Liguria Tirrenica Estratto Dal Rapporto Nazionale VAPI Con Aggiornamenti*; Technical Report; CNR-IRPI, GNDCI: Milano, Italy, 2001.
51. Ballarin, C.; Majone, U.; Mignosa, P.; Tomirotti, M. Una metodologia di stima indiretta degli idrogrammi sintetici per il progetto di opere di difesa idraulica del territorio. *L'Acqua* **2001**, *3*, 9–16.
52. Bacchi, B.; Brath, A.; Kottegoda, N.T. Analysis of the relationships between flood peaks and flood volumes based on crossing properties of river flow processes. *Water Resour. Res.* **1992**, *28*, 2773–2782. [CrossRef]
53. Franchini, M.; Galeati, G. Comparative analysis of some methods for deriving the expected flood reduction curve in the frequency domain. *Hydrol. Earth Syst. Sci.* **2000**, *4*, 155–172. [CrossRef]
54. Singh, V.P. *Computer Models of Watershed Hydrology*; Water Resources Publications: Littleton, Colorado, 1995; ISBN 0918334918.
55. Tomirotti, M.; Mignosa, P. A methodology to derive Synthetic Design Hydrographs for river flood management. *J. Hydrol.* **2017**, *555*, 736–743. [CrossRef]
56. Toro, E.F. *Shock-Capturing Methods for Free-Surface Shallow Flows*; Wiley: Hoboken, NJ, USA, 2001; ISBN 0471987662.
57. Liang, Q.; Marche, F. Numerical resolution of well-balanced shallow water equations with complex source terms. *Adv. Water Resour.* **2009**, *32*, 873–884. [CrossRef]
58. Dazzi, S.; Vacondio, R.; Dal Palù, A.; Mignosa, P. A local time stepping algorithm for GPU-accelerated 2D shallow water models. *Adv. Water Resour.* **2018**, *111*, 274–288. [CrossRef]
59. Dazzi, S.; Vacondio, R.; Mignosa, P. Integration of a Levee Breach Erosion Model in a GPU-Accelerated 2D Shallow Water Equations Code. *Water Resour. Res.* **2019**, *55*, 682–702. [CrossRef]
60. Mignosa, P.; Vacondio, R.; Aureli, F.; Dazzi, S.; Ferrari, A.; Prost, F. High resolution 2D modelling of rapidly varying flows: Some case studies. *Ital. J. Eng. Geol. Environ.* **2018**, *2018*, 143–160. [CrossRef]
61. Dazzi, S.; Vacondio, R.; Ferrari, A.; D'Oria, M.; Mignosa, P. Flood Simulation in Urban Areas Obtained by GPU-Accelerated 2D Shallow Water Model with Internal Boundary Conditions. In Proceedings of the Riverflow 2020, Delft, The Netherlands, 7–10 July 2020; Uijttewaal, W., J. Franca, M., Valero, D., Chavarrias, V., Ylla Arbós, C., Schielen, R., Crosato, A., Eds.; CRC Press: Boca Raton, FL, USA, 2020; pp. 1130–1138.
62. Aureli, F.; Prost, F.; Vacondio, R.; Dazzi, S.; Ferrari, A. A GPU-Accelerated Shallow-Water Scheme for Surface Runoff Simulations. *Water* **2020**, *12*, 637. [CrossRef]

63. Bradley, J.N. *Hydraulics of Bridge Waterways*; US Department of Transportation, Federal Highway Administration: Washington DC, USA, 1973.
64. Schubert, J.E.; Sanders, B.F. Building treatments for urban flood inundation models and implications for predictive skill and modeling efficiency. *Adv. Water Resour.* **2012**, *41*, 49–64. [CrossRef]
65. Chow, V.T. *Open-Channel Hydraulics*; McGraw-Hill Book Company: New York, NY, USA, 1959; ISBN 0-07-085906-X.
66. Valiani, A.; Caleffi, V.; Zanni, A. Case Study: Malpasset Dam-Break Simulation using a Two-Dimensional Finite Volume Method. *J. Hydraul. Eng.* **2002**, *128*, 460–472. [CrossRef]
67. Petaccia, G.; Natale, L. 1935 Sella Zerbino Dam-Break Case Revisited: A New Hydrologic and Hydraulic Analysis. *J. Hydraul. Eng.* **2020**, *146*, 05020005. [CrossRef]
68. Pilotti, M.; Maranzoni, A.; Tomirotti, M.; Valerio, G. 1923 Gleno Dam Break: Case Study and Numerical Modeling. *J. Hydraul. Eng.* **2011**, *137*, 480–492. [CrossRef]
69. Chow, V.T.; Maidment, D.R.; Mays, L.W. *Applied Hydrology*; McGraw-Hill Book Company: New York, NY, USA, 1988; ISBN 0-07-100174-3.
70. Bizzi, S.; Dinh, Q.; Bernardi, D.; Denaro, S.; Schippa, L.; Soncini-Sessa, R. On the control of riverbed incision induced by run-of-river power plant. *Water Resour. Res.* **2015**, *51*, 5023–5040. [CrossRef]
71. Aureli, F.; Ziveri, C.; Maranzoni, A.; Mignosa, P. Fully-2D and quasi-2D modeling of flooding scenarios due to embankment failure. In *River Flow 2006*; Taylor and Francis Group: London, UK, 2006; pp. 6–8, ISBN 0-415-40815-6.

Article

Flood Early Warning Systems Using Machine Learning Techniques: The Case of the Tomebamba Catchment at the Southern Andes of Ecuador

Paul Muñoz [1,2,*], Johanna Orellana-Alvear [1,2], Jörg Bendix [3], Jan Feyen [4] and Rolando Célleri [1,2]

[1] Departamento de Recursos Hídricos y Ciencias Ambientales, Universidad de Cuenca, Cuenca 010150, Ecuador; johanna.orellana@ucuenca.edu.ec (J.O.-A.); rolando.celleri@ucuenca.edu.ec (R.C.)
[2] Facultad de Ingeniería, Universidad de Cuenca, Cuenca 010150, Ecuador
[3] Laboratory for Climatology and Remote Sensing, Faculty of Geography, University of Marburg, 35032 Marburg, Germany; bendix@mailer.uni-marburg.de
[4] Faculty of Bioscience Engineering, Catholic University of Leuven, 3001 Leuven, Belgium; jan.feyen@kuleuven.be
* Correspondence: paul.munozp@ucuenca.edu.ec

Abstract: Worldwide, machine learning (ML) is increasingly being used for developing flood early warning systems (FEWSs). However, previous studies have not focused on establishing a methodology for determining the most efficient ML technique. We assessed FEWSs with three river states, *No-alert*, *Pre-alert* and *Alert* for flooding, for lead times between 1 to 12 h using the most common ML techniques, such as multi-layer perceptron (MLP), logistic regression (LR), K-nearest neighbors (KNN), naive Bayes (NB), and random forest (RF). The Tomebamba catchment in the tropical Andes of Ecuador was selected as a case study. For all lead times, MLP models achieve the highest performance followed by LR, with $f1$-macro (*log-loss*) scores of 0.82 (0.09) and 0.46 (0.20) for the 1 h and 12 h cases, respectively. The ranking was highly variable for the remaining ML techniques. According to the g-mean, LR models correctly forecast and show more stability at all states, while the MLP models perform better in the *Pre-alert* and *Alert* states. The proposed methodology for selecting the optimal ML technique for a FEWS can be extrapolated to other case studies. Future efforts are recommended to enhance the input data representation and develop communication applications to boost the awareness of society of floods.

Keywords: flood early warning; forecasting; hydrological extremes; machine learning; Andes

Citation: Muñoz, P.; Orellana-Alvear, J.; Bendix, J.; Feyen, J.; Célleri, R. Flood Early Warning Systems Using Machine Learning Techniques: The Case of the Tomebamba Catchment at the Southern Andes of Ecuador. *Hydrology* **2021**, *8*, 183. https://doi.org/10.3390/hydrology8040183

Academic Editors: Marina Iosub and Andrei Enea

Received: 25 November 2021
Accepted: 13 December 2021
Published: 16 December 2021

1. Introduction

Flooding is the most common natural hazard and results worldwide in the most damaging disasters [1–4]. Recent studies associate the increasing frequency and severity of flood events with a change in land use (e.g., deforestation and urbanization) and climate [2,5–7]. This particularly holds for the tropical Andes region, where complex hydro-meteorological conditions result in the occurrence of intense and patchy rainfall events [8–10].

According to the flood generation mechanism, floods can be classified into long- and short-rain floods [11,12]. A key for building resilience to short-rain floods is to anticipate in a timely way the event, in order to gain time for better preparedness. The response time between a rainfall event and its associated flood depends on the catchment properties and might vary from minutes to hours [13]. In this study special attention is given to flash-floods, which are floods that develop less than 6 h after a heavy rainfall with little or no forecast lead time [14].

Flood anticipation can be achieved through the development of a flood early warning system (FEWS). FEWSs have proved to be cost-efficient solutions for life preservation, damage mitigation, and resilience enhancement [15–18]. However, although crucial, flood forecasting remains a major challenge in mountainous regions due to the difficulty to effectively record the aerial distribution of precipitation due to the sparse density of the monitoring network and the absence of high-tech equipment by budget constraints [8,9].

To date, there has been no report of any operational FEWS in the Andean region for scales other than continental [17,19,20]. An alternative attempt in Peru aimed to derive daily maps of potential floods based on the spatial cumulated precipitation in past days [21]. Other endeavors in Ecuador and Bolivia focused on the monitoring of the runoff in the upper parts of the catchment to predict the likelihood of flood events in the downstream basin area [19,22]. However, such attempts are unsatisfactory as countermeasures against floods and especially flash-floods, where it is required to have reliable and accurate forecasts with lead times shorter than the response time between the farthest precipitation station and runoff control point.

There are two paradigms that drive the modeling of the precipitation-runoff response. First, the physically-based paradigm includes knowledge of the physical processes by using physical process equations [23]. This approach requires extensive ground data and, in consequence, intensive computation that hinders the temporal forecast window [24]. Moreover, it is argued that physically based models are inappropriate for real-time or short-term flood forecasting due to the inherent uncertainty of river-catchment dynamics and over-parametrization of this type of model [25]. The second data-driven paradigm assumes floods as stochastic processes with an occurrence distribution probability derived from historical data. Here, the idea is to exploit relevant input information (e.g., precipitation, past runoff) to find relations to the target variable (i.e., runoff) without requiring knowledge about the underlying physical processes. Among the traditional data-driven approaches, statistical modeling has proven to be unsuitable for short-term prediction due to lack of accuracy, complexity, model robustness, and even computational costs [24]. Previous encouraged the use of advanced data-driven models, e.g., machine learning (ML), to overcome the aforementioned shortcomings [7,24,26,27]. Particularly during the last decade, ML approaches have gained increasing popularity among hydrologists [24].

Different ML strategies for flood forecasting are implemented, generating either quantitative or qualitative runoff forecasts [18,28–38]. Qualitative forecasting consists of classifying floods into distinct categories or river states according to their severity (i.e., runoff magnitude), and use this as a base for flood class prediction [30,37,39]. The advantage of developing a FEWS is the possibility to generate a semaphore-like warning system that is easy to understand by decision-makers and the public (non-hydrologists). The challenge of FEWSs is the selection of the most optimal ML technique to obtain reliable and accurate forecasts with sufficient lead time for decision making. To date, the problem has received scant attention in the research literature, and as far as our knowledge extends no previous work examined and compared the potential and efficacy of different ML techniques for flood forecasting.

The present study compares the performance of five ML classification techniques for short-rain flood forecasting with special attention to flash floods. ML models were developed for a medium-size mountain catchment, the Tomebamba basin located in the tropical Andes of Ecuador. The ML models were tested with respect to their capacity to forecast three flood warning stages (*No-alert*, *Pre-alert* and *Alert*) for varying forecast lead times of 1, 4, and 6 h (flash-floods), but also 8 and 12 h to further test whether the lead time can be satisfactorily extended without losing the models' operational value.

This paper has been organized into four sections. The first section establishes the methodological framework for developing a FEWSs using ML techniques. It will then go on to describe the performance metrics used for a proper efficiency assessment. The second section presents the findings of the research following the same structure as the methodological section. Finally, the third and fourth sections presents the discussion and a summary of the main conclusions of the study, respectively.

2. Materials and Methods

2.1. Study Area and Dataset

The study area comprises the Tomebamba catchment delineated upstream of the Matadero-Sayausí hydrological station of the Tomebamba river (Figure 1), where the river enters the city. The Tomebamba is a tropical mountain catchment located in the southeastern flank of the Western Andean Cordillera, draining to the Amazon River. The drainage area of the catchment is approximately 300 km^2, spanning from 2800 to 4100 m above sea level (m a.s.l.). Like many other mountain catchments of the region, it is primarily covered by a páramo ecosystem, which is known for its important water regulation function [8].

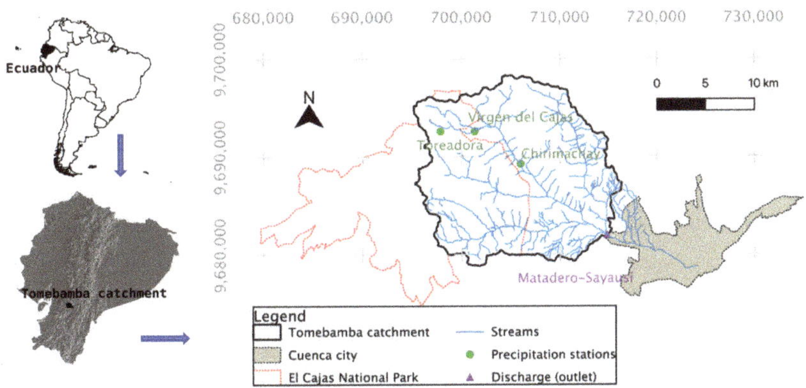

Figure 1. The Tomebamba catchment located at the Tropical Andean Cordillera of Ecuador, South America (UTM coordinates).

The Tomebamba river plays a crucial role as a drinking water source for the city of Cuenca (between 25% to 30% of the demand). Other important water users are agricultural and industrial activities. Cuenca, which is the third-largest city of Ecuador (around 0.6 million inhabitants), is crossed by four rivers that annually flood parts of the city, causing human and significant economic losses.

The local water utility, the Municipal Public Company of Telecommunications, Water, Sewerage and Sanitation of Cuenca (ETAPA-EP), defined three flood alert levels associated with the Matadero-Sayausí station for floods originating in the Tomebamba catchment: (i) *No-alert* of flood occurs when the measured runoff is less than 30 m^3/s, (ii) *Pre-alert* when runoff is between 30 and 50 m^3/s, and (iii) the flood *Alert* is triggered when discharge exceeds 50 m^3/s. With these definitions, and as shown in Figure 2, the discharge label for the *No-alert* class represents the majority of the data, whereas the *Pre-alert* and *Alert* classes comprise the minority yet the most dangerous classes.

To develop and operate forecasting models, we use data of two variables: precipitation in the catchment area and river discharge at a river gauge. For both variables, the available dataset comprises 4 years of concurrent hourly time series, from Jan/2015 to Jan/2019 (Figure 2). Precipitation information was derived from three tipping-bucket rain gauges: Toreadora (3955 m a.s.l.), Virgen (3626 m a.s.l.), and Chirimachay (3298 m a.s.l.) installed within the catchment and along its altitudinal gradient. Whereas for discharge, we used data of the Matadero-Sayausí station (2693 m a.s.l., Figure 1). To develop the ML modes, we split the dataset into training and test subsets. The training period ran from 2015 to 2017, whereas 2018 was used as the model testing phase.

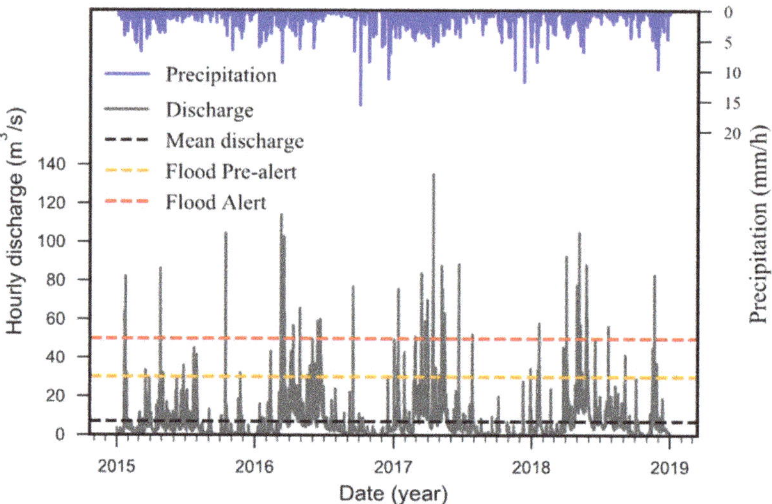

Figure 2. Time series of precipitation (Toreadora) and discharge (Matadero-Sayausí). Horizontal dashed lines indicate the mean runoff and the currently employed flood alert levels for labeling the *Pre-alert* and *Alert* flood warnings classes.

2.2. Machine Learning (ML) Methods for Classification of Flood Alert Levels

ML classification algorithms can be grouped in terms of their functionality. According to Mosavi et al. (2018), five of the worldwide most-popular statistical method groups are commonly used for short-term flood prediction (extreme runoff), and include:

i. Regression algorithms modeling the relationships between variables (e.g., logistic regression, linear regression, multivariate adaptive regression splines, etc.) [18,40].
ii. Instance-based algorithms that rely on memory-based learning, representing a decision problem fed with data for training (e.g., K-nearest neighbor, learning vector quantification, locally weighted learning, etc.) [30].
iii. Decision tree algorithms, which progressively divide the whole data set into subsets based on certain feature values, until all target variables are grouped into one category (e.g., classification and regression tree, M5, random forest, etc.) [18,28,30,31,37].
iv. Bayesian algorithms based on Bayes' theorem on conditional probability (e.g., naive Bayes, Bayesian network, Gaussian naïve Bayes, etc.) [18,31,35].
v. Neural Network algorithms inspired by biological neural networks convert input(s) to output(s) through specified transient states that enable the model to learn in a sophisticated way (e.g., perceptron, multi-layer perceptron, radial basis function network, etc.) [18,31,36].

For this study, we selected five ML algorithms, one from each group, respectively, a logistic regression, K-nearest neighbor, random forest, naive Bayes, and a multi-layer perceptron.

2.2.1. Logistic Regression

Logistic Regression (LR) is a discriminative model, modeling the decision boundary between classes. In a first instance, linear regressions are applied to find existent relationships between model features. Thereafter, the probability (conditional) of belonging to a class is identified using a logistic (sigmoid) function that effectively deals with outliers (binary classification). From these probabilities, the LR classifies, with regularization, the dependent variables into any of the created classes. However, for multiclass classification problems are all binary classification possibilities considered, it is *No-alert* vs. *Pre-alert*, *No-alert* vs. *Alert*, and *Pre-alert* vs. *Alert*. Finally, the solution is the classification with the maximum probability (multinomial LR) using the *softmax* function Equation (1). With this function is the predicted probability of each class defined [41]. The calculated probability for each class is positive with the logistic function and normalized across all classes.

$$softmax(z)_i = \frac{e^{z_i}}{\sum_{l=1}^{k} e^{z_l}} \tag{1}$$

where z_i is the ith input of the *softmax* function, corresponding to class i from the k number of classes.

2.2.2. K-Nearest Neighbors

K-Nearest Neighbors (KNN) is a non-parametric statistical pattern recognition algorithm, for which no theoretical or analytical background exist but an intuitive statistical procedure (memory-based learning) for the classification. KNN classifies unseen data based on a similarity measure such as a distance function (e.g., Euclidean, Manhattan, Chebyshev, Hamming, etc.). The use of multiple neighbors instead of only one is recommended to avoid the wrong delineation of class boundaries caused by noisy features. In the end, the majority vote of the nearest neighbors (see the formulation in [41]) determines the classification decision. The number of nearest neighbors can be optimized to reach a global minimum avoiding longer computation times, and the influence of class size. The major advantage of the KNN is its simplicity. However, the drawback is that KNN is memory intensive, all training data must be stored and compared when added information is to be evaluated.

2.2.3. Random Forest

Random Forest (RF) is a supervised ML algorithm that ensembles a multitude of decorrelated decision trees (DTs) voting for the most popular class (classification). In practice, a DT (particular model) is a hierarchical analysis based on a set of conditions consecutively applied to a dataset. To assure decorrelation, the RF algorithm applies a bagging technique for a growing DT from different randomly resampled training subsets obtained from the original dataset. Each DT provides an independent output (class) of the phenomenon of interest (i.e., runoff), contrary to numerical labels for regression applications. The popularity of RF is due to the possibility to perform random subsampling and bootstrapping which minimizes biased classification [42]. An extended description of the RF functioning is available in [43,44].

The predicted class probabilities of an input sample are calculated as the mean predicted class probabilities of the trees in the forest. For a single tree, the class probability is computed as the fraction of samples of the same class in a leaf. However, it is well-known that the calculated training frequencies are not accurate conditional probability estimates due to the high bias and variance of the frequencies [45]. This deficiency can be resolved by controlling the minimum number of samples required at a leaf node, with the objective to induce a smoothing effect, and to obtain statistically reliable probability estimates.

2.2.4. Naïve Bayes

Naïve Bayes (NB) is a classification method based on Bayes' theorem with the "naive" assumption that there is no dependence between features in a class, even if there is dependence [46]. Bayes' theorem can be expressed as:

$$P(y|X) = \frac{P(X|y)\,P(y)}{P(X)} \tag{2}$$

where $P(A|B)$ is the probability of y (hypothesis) happening, given the occurrence of X (features), and X can be defined as $X = x_1, x_2, \ldots, x_n$. Bayes' theorem can be written as:

$$P(y|x_1,\, x_2, \ldots, x_n) = \frac{P(x_1|y)\,P(x_2|y)\ldots\,P(x_n|y)\,P(y)}{P(x_1)\,P(x_2)\ldots\,P(x_n)} \tag{3}$$

There are different NB classifiers depending on the assumption of the distribution of $P(x_i \,|y)$. In this matter, the study of Zhang [46] proved the optimality of NB under the Gaussian distribution even when the assumption of conditional independence is violated (real application cases). Additionally, for multiclass problems, the outcome of the algorithm is the class with the maximum probability. For the Gaussian NB algorithm no parameters have to be tuned.

2.2.5. Multi-Layer Perceptron

The Multi-Layer Perceptron (MLP) is a class of feedforward artificial neural networks (ANN). A perceptron is a linear classifier that separates an input into two categories with a straight line and produces a single outcome. Input is a feature vector multiplied by specific weights and added to a bias. Contrary to a single-layer case, the MLP can approximate non-linear functions using additional so-called hidden layers. Prediction of probabilities of belonging to any class is calculated through the *softmax* function. The MLP consists of multiple neurons in fully connected multiple layers. Determination of the number of neurons in the layers with a trial-and-error approach remains widely used [47]. Neurons in the first layer correspond to the input data, whereas all other nodes relate inputs to outputs by using linear combinations with certain weights and biases together with an activation function. To measure the performance of the MLP, the logistic loss function is defined with the limited memory Broyden–Fletcher–Goldfarb–Shanno (L-BFGS) method as the optimizer for training the network. A detailed and comprehensive description of ANN can be found in [48].

2.3. *Methodology*

Figure 3 depicts schematic the methodology followed in this study. The complete dataset for the study consists, as mentioned before, of precipitation and labeled discharge time-series (see Figure 2). The dataset was split in two groups, respectively, for training and testing purposes, and training and test feature spaces were composed for each lead time for the tasks of model hyperparameterization and model assessment. This procedure is repeated for each of the ML techniques studied. Finally, the ranking of the performance quality of all ML methods for every lead time, based on performance metrics and a statistical significance test, were determined.

2.3.1. Feature Space Composition

For each lead time, we used single training and testing feature spaces for all ML techniques. A feature space is composed by features (predictors) coming from two variables: precipitation and discharge. The process of feature space composition starts by defining a specific number of precipitation and discharge features (present time and past hourly lags) according to statistical analyses relying on Pearson's cross-, auto and partial-auto-correlation functions [49]. The number of lags from each station was selected by setting up a correlation threshold of 0.2 [28].

Similarly, for discharge, we used several features coming from past time slots of discharge selected for the analysis. It is worth noting that the number of discharge features triples since we replace each discharge feature with three features (one per flood warning class) in a process known as one-hot-encoding or binary encoding. Therefore, each created feature denotes 0 or 1 when the correspondent alarm stage is false or true, respectively. Finally, we performed a feature standardization process before the computation stage of the KNN, LR, NB, and NN algorithms. Standardization was achieved by subtracting the mean and scaling it to unit variance, resulting in a distribution with a standard deviation equal to 1 and a mean equal to 0.

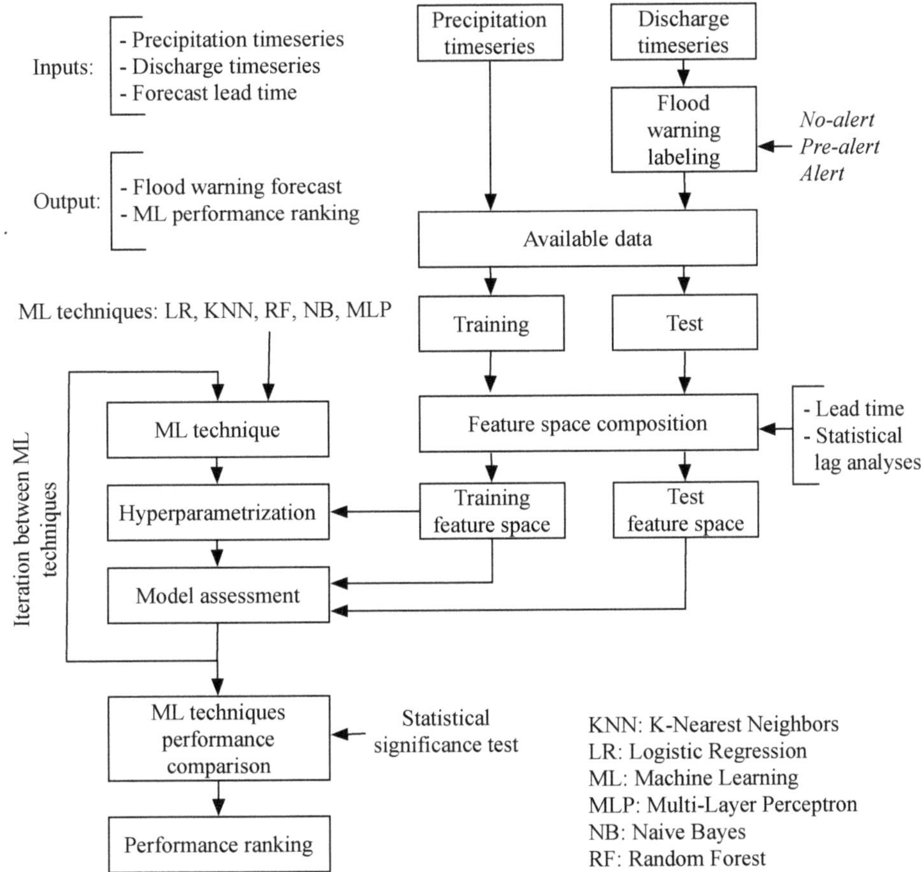

Figure 3. Work schedule for the development and evaluation of the machine learning (ML) flood forecasting models.

2.3.2. Model Hyperparameterization

After the composition of the feature space the optimal architectures for each ML forecasting model, and for each lead time was set up. The optimal architectures were defined by the combination of hyperparameters under the concept of balance between accuracy, and computational cost, and speed. However, finding optimal architectures requires an exhaustive search of all combinations of hyperparameters. To overcome this issue, we relied on the randomized grid search (RGS) with a 10-fold cross-validation scheme. The RGS procedure randomly explores the search space for discretized continuous hyperparameters based on a cross-validation evaluation. Moreover, we selected the f1-macro score (see Section 2.3.4) as objective function.

2.3.3. Principal Component Analysis

ML applications require in general the analysis of high-dimension and complex data, involving substantial amounts of memory and computational costs. Reduction of the dimensionality was realized through the application of principal component analysis (PCA) enabling exclusion of correlating features that do not add information to the model. PCA was applied after feature scaling and normalization.

This method enables finding the dimension of maximum variance and the reduction of the feature space to that dimension so that the model performance remains as intact as possible when compared to performance with the full feature space. But considering that each ML technique assimilates data differently, we did not define the number of principal components to keep a fixed threshold of variance explanation (e.g., 80–90%), but performed an exploratory analysis to evaluate its influence on each model. As such, the number of PCAs was treated as an additional hyperparameter, and we optimized the number of principal components for each specific model (lead time and ML technique) with the objective to find the best possible model for each case.

All ML techniques and the RGS procedure were implemented through the scikit-learn package for ML in Python® [50]. Table 1 presents the relevant hyperparameters for each ML technique and their search space for tuning [38]. We employed default values for the hyperparameters which are depicted in Table 1.

Table 1. Model hyperparameters and their ranges/possibilities for tuning.

ML Technique	Hyperparameters				
LR	C 0.001–1000	penalty {'l1', 'l2'}			
KNN	neighbor's 3–75	weights {'uniform', 'distance'}	metric {'euclidean', 'manhattan', 'minkowski'}	algorithm {'auto','ball_tree', 'kd_tree','brute'}	
RF	estimator's 50–1000	max_features {'auto', 'sqrt', 'log2'}	hadeeth 50–1000	min_samples_leaf 1–500	min_samples_split 1–500
MLP	solver {'lbfgs'}	max_iter 10–5000	alpha 1×10^{-9}–0.1	hidden_layers 1–16	

2.3.4. Model Performance Evaluation

Forecasting hydrological extremes such as floods turns into an imbalanced classification problem, and becomes even more complex when the interest lies in the minority class of the data (flood alert). This is because most ML classification algorithms focus on the minimization of the overall error rate, it is the incorrect classification of the majority class [51]. Resampling the class distribution of the data for obtaining an equal number of samples per class is one solution. In this study, we used another approach that relies on training ML models with the assumption of imbalanced data. The approach we used penalizes mistakes in samples belonging to the minority classes rather than under-sampling or over-sampling data. In practice, this implies that for a given metric efficiency, the overall score is the result of averaging each performance metric (for each class) multiplied by its corresponding weight factor. According to the class frequencies the weight factors for each class were calculated (inversely proportional), using Equation (4).

$$w_i = \frac{N}{C\, n_j} \quad (4)$$

where w_i is the weight of class i, N is the total number of observations, C is the number of classes, and n_j the number of observations in class i. This implies that higher weights will be obtained for minority classes.

Performance Metrics

The metrics for the performance assessment were derived from the well-known confusion matrix, especially suitable for imbalanced datasets and multiclass problems, and are respectively the f1 score, the geometric mean, and the logistic regression loss score [51–56]. Since neither of the metrics is adequate it is suggested to use a compendium of metrics to properly explain the performance of the model. In addition, those metrics complement each other.

f1 Score

The f score is a metric that relies on precision and recall, which is an effective metric for imbalanced problems. When the f score as a weighted harmonic mean, we name this score f1. The latter score can be calculated with Equation (5).

$$\text{f1 score} = \frac{2 \times \text{Precision} \times \text{Recall}}{(\text{Precision} + \text{Recall})} \quad (5)$$

where precision and recall are defined with the following equations:

$$\text{Precision} = \frac{\text{TP}}{\text{TP} + \text{FP}} \quad (6)$$

$$\text{Recall} = \frac{\text{TP}}{\text{TP} + \text{FN}} \quad (7)$$

where TP stands for true positives, FP for false positives, and FN for false negatives.

The f1 score ranges from 0 to 1, indicating perfect precision and recall. The advantage of using the f1 score compared to the arithmetic or geometric mean is that it penalizes models most when either the precision or recall is low. However, classifying a *No-Alert* flood warning as *Alert* might have a different impact on the decision-making than when the opposite occurs. This limitation scales up when there is an additional state, e.g., *Pre-alert*. Thus, the interpretation of the f1 score must be taken with care. For multiclass problems, the f1 score is commonly averaged across all classes, and is called the f1-macro score to indicate the overall model performance.

Geometric Mean

The geometric-mean (g-mean) measures simultaneously the balanced performance of TP and TN rates. This metric gives equal importance to the classification task of both the majority (*No-alert*) and minority (*Pre-alert* and *Alert*) classes. The g-mean is an evaluation measure that can be used to maximize accuracy to balance TP and TN examples at the same time with a good trade-off [53]. It can be calculated using Equation (8)

$$\text{G} - \text{mean} = \sqrt{(\text{TP}_{\text{rate}} \times \text{TN}_{\text{rate}})} \quad (8)$$

where TP_{rate} and TN_{rate} are defined by:

$$\text{TP}_{\text{rate}} = \text{Recall} \quad (9)$$

$$\text{TN}_{\text{rate}} = \frac{\text{TN}}{\text{TN} + \text{FP}} \quad (10)$$

The value of the g-mean metrics ranges from 0 to 1, where low values indicate deficient performance in the classification of the majority class even if the minority classes are correctly classified.

Logistic Regression Loss

The metric logistic regression loss (log-loss) measures the performance of a classification model when the input is a probability value between 0 and 1. It accounts for the uncertainty of the forecast based on how much it varies from the actual label. For multiclass classification, a separate log-loss is calculated for each class label (per observation), and the results are summed up. The log-loss score for multi-class problems is defined as:

$$Log\ loss = -\frac{1}{N}\sum_{i=1}^{N}\sum_{j=1}^{M} y_{ij}\log(p_{ij}) \quad (11)$$

where N is the number of samples, M the number of classes, y_{ij} equal to 1 when the observation belongs to class j; else 0, and p_{ij} is the predicted probability that the observation belongs to class j. Starting from 0 (best score), the log-loss magnitudes increase as the probability diverges from the actual label. It punishes worse errors more harshly to promote conservative predictions. For probabilities close to 1, the log-loss slowly decreases. However, as the predicted probability decreases, the log-loss increases rapidly.

Statistical Significance Test for Comparing Machine-Learning (ML) Algorithms

Although we can directly compare performance metrics of ML alternatives and claim to have found the best one based on the score, it is not certain whether the difference in metrics is real or the result of statistical chance. Different statistical frameworks are available allowing us to compare the performance of classification models (e.g., a difference of proportions, paired comparison, binomial test, etc.).

Among them, Raschka [57] recommends using the chi-squared test to quantify the likelihood of the samples of skill scores, being observed under the assumption that they have the same distributions. The assumption is known as the null hypothesis, and aims to prove whether there is a statistically significant difference between two models (error rates). If rejected, it can be concluded that any observed difference in performance metrics is due to a difference in the models and not due to statistical chance. In our study we used the chi-squared test to assess whether the difference in the observed proportions of the contingency tables of a pair of ML algorithms (for a given lead time) is significant.

For the model comparison, we defined the statistical significance of improvements/ degradations for all lead times (training and test subsets) under a value of 0.05 (chi-squared test). In all cases, the MLP model was used as the base model to which the other models were compared.

3. Results

This section presents the results of the flood forecasting models developed with the LR, KNN, RF, NB, and MLP techniques, and for lead times of 1, 4, 6, 8, and 12 h. For each model, we addressed the forecast of three flood warnings, *No-alert*, *Pre-alert* and *Alert*. First, we present the results of the feature space composition process, taking the 1 h lead time case as an example. Then, we show the results of the hyperparameterization for all models, followed by an evaluation and ranking of the performance of the ML techniques.

3.1. Feature Space Composition

Figures 4 and 5 show the results of the discharge and precipitation lag analyses for the flood forecasting model 1-h before the flood would occur. Figure 4a depicts the discharge autocorrelation function (ACF) and the corresponding 95% confidence interval from lag 1 up to 600 (h). We found a significant correlation up to a lag of 280 h (maximum correlation at the first lag) and, thereafter, the correlation fell within the confidence band. On the other hand, Figure 4b presents the discharge partial-autocorrelation function (PACF) and its 95% confidence band from lag 1 to 30 h. We found a significant correlation up to lag 8 h (first lags outside the confidence band). As a result, based on the interpretation of the ACF and PACF analyses, and according to Muñoz et al. [28] we decided to include 8 discharge lags (hours) for the case of 1 h flood forecasting in the Tomebamba catchment.

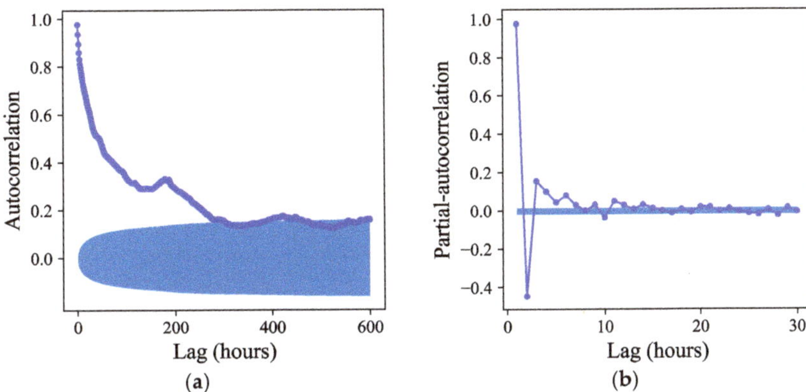

Figure 4. (**a**) Autocorrelation function (ACF) and (**b**) partial-autocorrelation function (PACF) of the Matadero-Sayausí (Tomebamba catchment) discharge series. The blue hatch indicates in each case the correspondent 95% confidence interval.

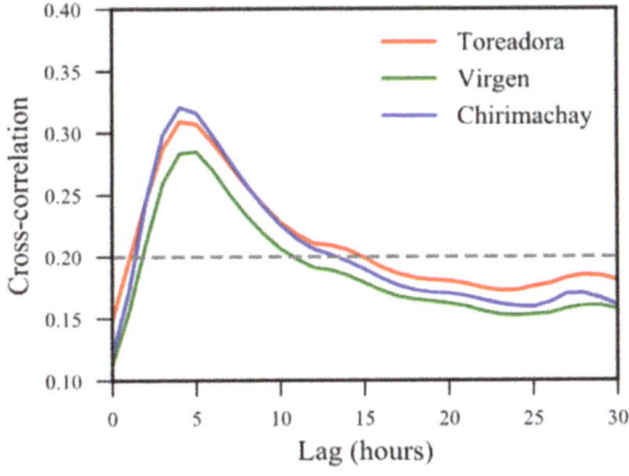

Figure 5. Pearson's cross-correlation comparison between the Toreadora (3955 m a.s.l.), Virgen (3626 m a.s.l.), and Chirimachay (3298 m a.s.l.) precipitation stations and the Matadero-Sayausí discharge series. Note the blue horizontal line at a fixed correlation of 0.2 for determining past lags.

Figure 5 plots the Pearson's cross-correlation between the precipitation at each rainfall station and the discharge at the Matadero-Sayausí stream gauging station. For all stations, we found a maximum correlation at lag 4 (maximum 0.32 for Chirimachay). With the fixed correlation threshold of 0.2, we included 11, 14, and 15 lags for Virgen, Chirimachay, and Toreadora stations, respectively.

Similarly, the same procedure was applied for the remaining lead times (i.e., 4, 6, 8, and 12 h). In Table 2, we present the input data composition and the resulting total number of features obtained from the lag analyses for each forecasting model. For instance, for the 1 h case, the total number of features in the feature space equals 67, from which 43 are derived from precipitation (past lags and one feature from present time for each station), and 24 from discharge (one-hot-encoding).

Table 2. Input data composition (number of features) for all ML models of the Tomebamba catchment.

	Discharge Lags * (h)	Precipitation Lags (h)			
Lead Time (h)	Matadero-Sayausí	Toreadora	Chirimachay	Virgen	Number of Features
1	8	15	14	11	67
4	12	18	17	14	88
6	14	20	19	16	100
8	16	22	21	18	112
12	20	26	25	22	136

* Note that each discharge feature triples (three flood warning classes) after a one-hot-encoding process.

3.2. *Model Hyperparameterization*

The results of the hyperparameterization including the number of PCA components employed for achieving the best model efficiencies are presented in Table 3. No evident relation between the number of principal components and the ML technique nor the lead time was found. In fact, for some models we found differences in the f1-macro score lower than 0.01 for a low and high number of principal components. See for instance the case of the KNN models where the optimal number of components significantly decayed for lead times greater than 4 h. For the 1 h lead time, 96% of the components were used, whereas for the rest of the lead times only less than 8%.

Table 3. Model hyperparameters and number of principal components used for each specific model (ML technique and lead time).

ML Technique	Hyperparameter	Lead Time				
		1 h	4 h	6 h	8 h	12 h
LR	C	0.01	0.00001	0.0001	0.0001	0.001
	penalty	'12'	'12'	'12'	'12'	'12'
	PCA_components *	58	62	78	75	51
KNN	n_neighbors	15	15	23	33	55
	weights	'uniform'	'uniform'	'uniform'	'uniform'	'uniform'
	metric	'minkowski'	'minkowski'	'minkowski'	'minkowski'	'minkowski'
	Algorithm	'auto'	'auto'	'auto'	'auto'	'auto'
	PCA_components *	64	6	6	6	4
RF	n_estimators	700	700	700	700	800
	max_features	'sqrt'	'auto'	auto	'log2'	'auto'
	max_depth	350	350	350	350	300
	min_samples_leaf	450	450	480	480	450
	min_samples_split	10	5	5	2	4
	PCA_components *	66	79	90	45	78
NB	PCA_components *	63	64	87	89	15
MLP	solver	'lbfgs'	'lbfgs'	'lbfgs'	'lbfgs'	'lbfgs'
	max_iter	2000	2000	2000	2000	2000
	alpha	0.0001	0.0001	0.0001	0.0001	0.0001
	hidden_layers	2	3	2	2	4
	PCA_components *	63	51	64	76	4

* From the total number of features: 1 h = 67, 4 h = 88, 6 h = 100, 8 h = 112, 12 h = 136 features.

If we turn to the evolution of models' complexity with lead time (Table 3) more complex ML architectures are needed to forecast greater lead times. This is underpinned by the fact that the corresponding optimal models require for greater lead times a stronger regularization (lower values of C) for LR, a greater number of neighbors (n_neighbors) for KNN, more specific trees (lower values of min_samples_split) for RF and more hidden layers (hidden_layers) for MLP.

3.3. Model Performance Evaluation

As mentioned before, model performances calculated with the f1-score, g-mean, and log-loss score were weighted according to class frequencies. Table 4 presents the frequency distribution for the complete dataset, respectively, for the training and test subsets. Here, the dominance of the *No-alert* flood class is evident, with more than 95% of the samples in both subsets. With this information, the class weights for the training period were calculated as $w_{No-alert} = 0.01$, $w_{Pre-alert} = 0.55$ and $w_{Alert} = 0.51$.

Table 4. The number of samples and relative percentage for the entire dataset and the training and test subsets.

Warning	Complete	Training	Test
No-alert	32,596 (96.1%)	24,890 (96.2%)	7706 (95.7%)
Pre-alert	720 (2.1%)	473 (1.8%)	247 (3.1%)
Alert	609 (1.8%)	(2.0%)	100 (1.2%)

The results of the model performance evaluation for all ML models and lead times (test subset) are summarized in Table 5. We proved for all models that the differences in performance metrics for a given lead time were due to the difference in the ML techniques rather than to the statistical chance. As expected, ML models' ability to forecast floods decreased for a longer lead time. For instance, for the case of 1 h forecasting, we found a maximum f1-macro score of 0.88 (MLP) for the training and 0.82 (LR) for the test subset. Whereas, for the 12 h case, the maximum f1-macro score was 0.71 (MLP) for the training and 0.46 (MLP) for the test subset.

Table 5. Models' performance evaluation on the test subset. Bold fonts indicate the best performance for a given lead time.

Lead Time (h)	RF	KNN	LR	NB	MLP
		F1-macro score			
1	0.59	0.73	**0.82**	0.57	0.78
4	0.47	0.57	0.59	0.46	**0.62**
6	0.47	0.45	0.50	0.41	**0.51**
8	0.44	0.41	0.44	0.45	**0.51**
12	0.42	0.36	0.44	0.43	**0.46**
		G-mean			
1	0.86	0.77	**0.88**	0.81	0.83
4	0.75	0.63	**0.76**	0.73	0.71
6	0.70	0.56	**0.72**	0.68	0.62
8	**0.73**	0.53	0.67	0.62	0.62
12	0.69	0.50	**0.69**	0.64	0.56
		Log-loss score			
1	0.28	0.38	1.09	3.14	**0.09**
4	0.38	0.46	0.74	4.10	**0.11**
6	0.45	0.58	0.47	4.71	**0.14**
8	0.50	0.65	0.53	0.59	**0.16**
12	0.59	0.70	0.57	2.17	**0.20**

Note: All improvements and degradations are statistically significant.

The extensive hyperparameterization (RGS scheme) powered by 10-fold cross-validation served to assure robustness in all ML models and reduced overfitting. We found only a small difference between the performance values by using the training and the test subsets. For all models, maximum differences in performances were lower than 0.27 for the f1-macro score and 0.19 for the g-mean.

In general, for all lead times, the MLP technique obtained the highest f1-macro score, followed by the LR algorithm. This performance dominance was confirmed by the ranking of the models according to the log-loss score. The ranking of the remaining models was highly variable and, therefore, not conclusive. For instance, the results of the KNN models obtained the second-highest score for the training subset, but the lowest for the test subset, especially for longer lead times. This is because the KNN is a memory-based algorithm and therefore more sensitive to the inclusion of information different to the training subset in comparison to the remaining ML techniques. This can be noted in Table 4, where the training and test frequency distributions are different for the *Pre-alert* and *Alert* classes.

On the other hand, for the g-mean score, we obtained a different ranking of the methods. We found the highest scores for the LR algorithm, followed by the RF and the MLP models. Despite this behavior, the values of the g-mean were superior to the f1-macro scores for all lead times and subsets. This is because the f1 score relies on the harmonic mean. Therefore, the f1 score penalizes more a low precision or recall in comparison with a metric based on a geometric or arithmetic mean. Results of the g-mean served to identify that the LR is the most stable method in terms of correctly classifying both the majority (*No-alert*) and the minority (*Pre-alert* and *Alert*) flood warning classes, while the MLP technique could be used to focus on the minority (flood alert) classes.

To extend the last idea, we analyzed the individual f1 scores of each flood warning class. This unveils the ability of the model to forecast the main classes of interest, i.e., *Pre-alert* and *Alert*. Figure 6 presents the evolution of the f1-score of each ML algorithm at the corresponding lead time. We found that for all ML techniques, the *Alert* class is clearly the most difficult to forecast when the f1-macro score was selected as the metric for the hyperparameterization task. An additional exercise consisted in choosing the individual f1-score for the *Alert* class as the target for hyperparameterization of all models. However, although we obtained comparable results for the *Alert* class, the scores of the *Pre-alert* class had significantly deteriorated, even reaching scores near zero. The most interesting aspect in Figure 6 is that the most efficient and stable models across lead times (test subset) were the models based on MLP and LR techniques. It is also evident that for all forecasting models, a lack of robustness for the *Pre-alert* warning class was found, and there were major differences between the f1-scores for the training and test subsets. An explanation for this might be that the *Alert* class implies a *Pre-alert* warning class, but not the opposite. Consequently, this might mislead the learning process causing overfitting during training leading to poor performances when assessing unseen data during the test phase.

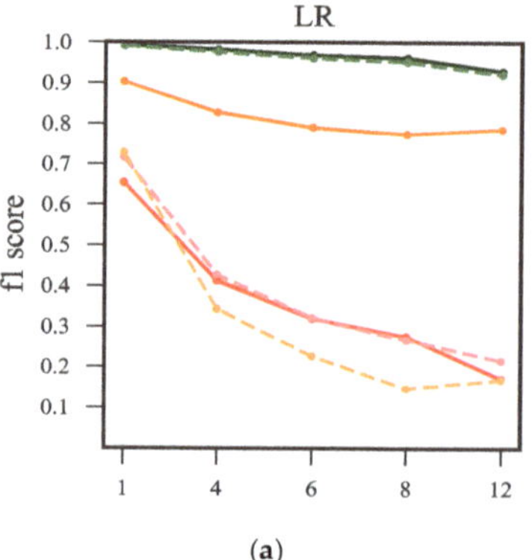

(a)

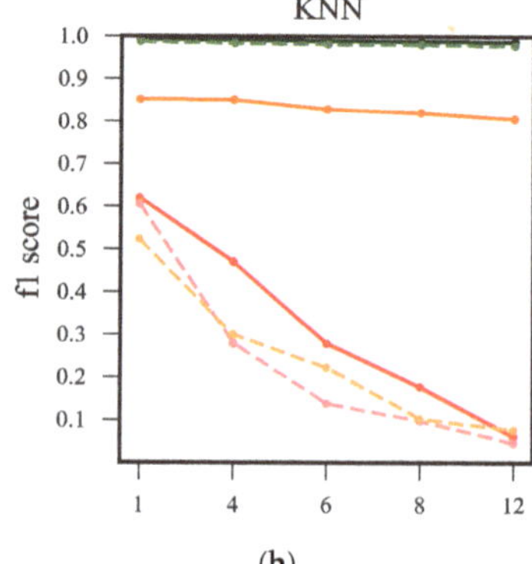

(b)

Figure 6. *Cont.*

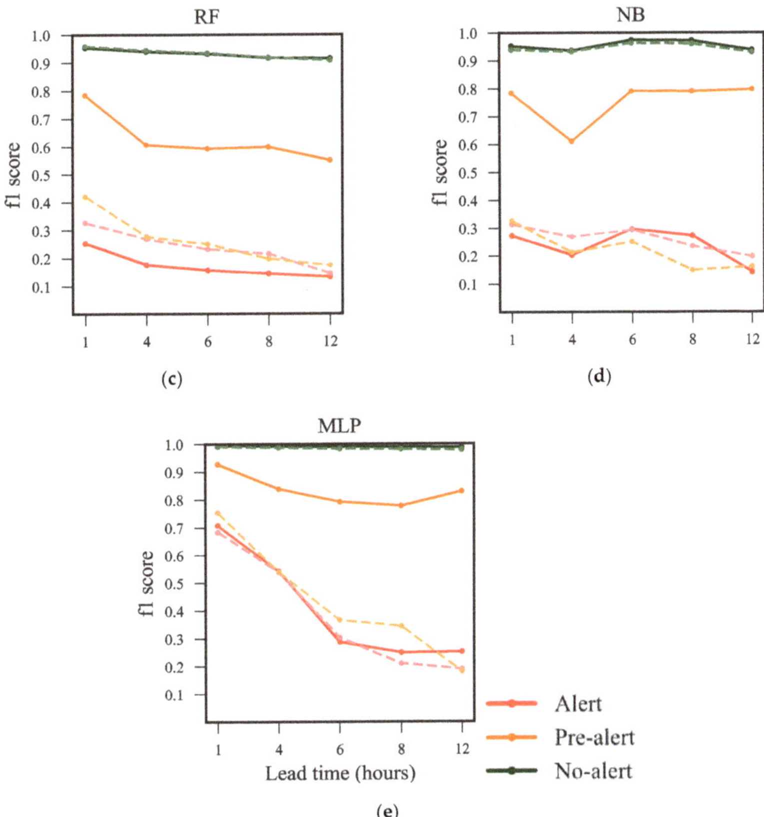

Figure 6. f1 scores per flood warning state (*No-alert*, *Pre-alert* and *Alert*) for all combinations of ML techniques across lead times. (**a**), Logistic Regression (**b**), K-Nearest Neighbors, (**c**) Random Forest, (**d**) Naïve Bayes, and (**e**) Multi-layer Perceptron. The brightest and dashed lines in each subfigure (color coding) represent the scores for the test subset.

Moreover, although we added a notion of class frequency distribution (weights) to the performance evaluation task, it can be noted that for all models, the majority class is most perfectly classified. This is because the *No-alert* class arises from low-to-medium discharge magnitudes. This eases and simplifies the learning process of the ML techniques since these magnitudes can be related to normal conditions (present time and past lags) of precipitation and discharge.

4. Discussion

In this study, we developed and evaluated five different FEWSs relying on the most common ML techniques for flood forecasting, and for short-term lead times of 1, 4, and 6 h for flash-floods, and 8 and 12 h to assess models' operational value for longer lead times. Historical runoff data were used to define and label the three flood warning scenarios to be forecasted (*No-alert*, *Pre-alert* and *Alert*). We constructed the feature space for the models according to the statistical analyses of precipitation and discharge data followed by a PCA analysis embedded in the hyperparameterization.

This was aimed to better exploit the learning algorithm of each ML technique. In terms of model assessment, we proposed an integral scheme based on the f1-score, the geometric mean, and the log-loss score to deal with data imbalance and multiclass characteristics. Finally, the assessment was complemented with a statistical analysis to provide a performance ranking between ML techniques. For all lead times, we obtained the best forecasts for both, the majority and minority classes from the models based on the LR, RF, and MLP techniques (g-mean). The two most suitable models for the dangerous warning classes (*Pre-Alert* and *Alert*) were the MLP and LR (f1 and log-loss scores). This finding has important implications for developing FEWSs since real-time applications must be capable of dealing with both the majority and minority classes. Therefore, it can be suggested that the most appropriate forecasting models are based on the MLP technique.

The results on the evolution of model performances across lead times suggest that the models are acceptable for lead times up to 6 h, i.e., the models are suitable for flash-flood applications in the Tomebamba catchment. For lead times greater than 6 h, we found a strong decay in model performance. In other words, the utility of the 8 and 12 h forecasting models is limited by the models' operational value. This is because, in the absence of rainfall forecasts, the assumption of future rain is solely based on runoff measurements at past and present times. This generates forecasts that are not accurate enough for horizons greater than the concentration-time of the catchment. The concentration-time of the Tomebamba catchment was estimated between 2 and 6 h according to the equations of Kirpich, Giandotti, Ven Te Chow, and Temez, respectively. A summary of the equations can be found in Almeida et al. [58]. This results in an additional performance decay for the 8 and 12 h cases in addition to the error in modeling.

The study of Furquim et al. [31] is comparable. These authors analyzed the performance of different ML classification algorithms for flash-flood nowcasting (3 h) in a river located in an urban area of Brazil. They found that models based on neural networks and decision trees outperformed those based on the NB technique. In addition, the study of Razali et al. [30] proved that decision tree-based algorithms perform better than KNN models, which agrees with our findings. However, such studies only evaluated the percentage of correctly classified instances which is a simplistic evaluation. Thus, we recommend a more integral assessment of model performances, like the one in the current study, which allows for better support in decision making.

Other studies related to quantitative forecasting revealed that neural network-based models usually outperform the remaining techniques proposed in our study [32–34]. Similarly, the study of Khalaf et al. [37] proved the superiority of the RF algorithm when compared to the bagging decision trees and HyperPipes classification algorithms. Thus, in certain cases, the use of less expensive techniques regarding the computational costs produces comparable results as in [36]; this is also the case in our short-rain and flash-flood flood classification problem.

As a further step, we propose the development of ensemble models for improving the performance results of individual models. This can be accomplished by combining the outcomes of the ML models with weights obtained, for instance, from the log-log scores. Another alternative that is becoming popular is the construction of hybrid models as a combination of ML algorithms for more accurate and efficient models [24,35,36]. Moreover, as stated by Solomatine and Xue [36], inaccuracies in forecasting floods are mainly due to data-related problems. In this regard, Muñoz et al. [9] reported a deficiency in precipitation-driven models due to rainfall heterogeneity in mountainous areas, where orographic rainfall formation occurs. In most cases, rainfall events are only partially captured by punctual measurement, and even the entire storm coverage can be missing.

In general precipitation-runoff models will reach at a certain point an effectiveness threshold that cannot be exceeded without incorporating new types of data such as soil moisture [59,60]. In humid areas, the rainfall–runoff relationship also depends on other variables such as evapotranspiration, soil moisture, and land use, which leads to significant spatial variations of water storage. However, these variables are difficult to measure or estimate.

5. Conclusions

- The current study set out to propose a methodology and integral evaluation framework for developing optimal short-rain flood warning forecasting models using ML classification techniques. The proposed analyses were applied to forecast three flood warnings, *No-alert*, *Pre-alert* and *Alert* for the Tomebamba catchment in the tropical Andes of Ecuador. For this, the five most common ML classification techniques for short-term flood forecasting were used. From the results, the following conclusion can be drawn: results related to model comparison are statistically significant. This is important because this is not usually performed in other studies and it validates the performance comparison and ranking hereby presented.
- For all lead times, the most suitable models for flood forecasting are based on the MLP followed by the LR techniques. From the integral evaluation (i.e., several performance metrics), we suggest LR models as the most efficient and stable option for classifying both the majority (*No-alert*) and the minority (*Pre-alert* and *Alert*) classes whereas we recommend MLP when the interest lies in the minority classes.
- The forecasting models we developed are robust. Differences in the averaged f1, g-mean and log-loss scores between training and test are consistent to all models. However, we limit the utility of the models for flash-flood applications (lead times up to 6 h). For longer lead times, we encourage improvement in precipitation representation, and even forecasting this variable for lead times longer than the concentration-time of the catchment.

A more detailed model assessment (individual f1 scores) demonstrated the difficulties of forecasting the *Pre-alert* and *Alert* flood warnings. This was evidenced when the hyperparameterization was driven for the optimization of the forecast for the alert class and this, however, did not improve the model performance of this specific class. This study can be extended with a deep exploration of the effect of input data composition, precipitation forecasting, and the feature engineering strategies for both the MLP and LR techniques. Feature engineering pursues the use of data representation strategies that could, for example, provide spatial and temporal information of the precipitation in the study area. This can be done by spatially discretizing precipitation in the catchments with the use of remotely sensed imagery. With this additional knowledge, it would be possible to improve the performance of the models hereby developed at longer lead times.

We recommend that future efforts should be put into applying the methodology and assessment framework proposed here in other tropical Andean catchments, and/or benchmarking the results obtained in this study with the outputs of physically based forecasting models. This was not possible for this study due to lack of data.

Finally, for FEWSs, the effectiveness of the models is strongly linked to the speed of communication to the public after a flood warning is triggered. Therefore, future efforts should focus on the development of a web portal and/or mobile application as a tool to boost the preparedness of society against floods that currently threaten people's lives, possessions, and environment in Cuenca and other comparable tropical Andean cities.

Author Contributions: Conceptualization, P.M. and R.C.; formal analysis, P.M.; funding acquisition, R.C.; investigation, P.M.; methodology, P.M. and J.O.-A.; project administration, R.C.; supervision, J.O.-A., J.B., J.F. and R.C.; writing—original draft, P.M.; writing—review and editing, J.O.-A., J.B., J.F. and R.C. All authors have read and agreed to the published version of the manuscript.

Funding: This research was funded by the project: "Desarrollo de modelos para pronóstico hidrológico a partir de datos de radar meteorológico en cuencas de montaña", funded by the Research Office of the University of Cuenca (DIUC), and the Empresa Pública Municipal de Telecomunicaciones, Agua Potable, Alcantarillado y Saneamiento de Cuenca (ETAPA-EP). Our thanks go to these institutions for their generous funding.

Institutional Review Board Statement: Not applicable.

Informed Consent Statement: Not applicable.

Data Availability Statement: All data, models, and code that support the findings of this study are available from the corresponding author upon reasonable request.

Acknowledgments: We acknowledge the Ministry of Environment of Ecuador (MAAE) for providing research permissions, and are grateful to the staff and students that contributed to the hydrometeorological monitoring. for experiments).

Conflicts of Interest: The authors declare no conflict of interest.

References

1. Stefanidis, S.; Stathis, D. Assessment of flood hazard based on natural and anthropogenic factors using analytic hierarchy process (AHP). *Nat. Hazards* **2013**, *68*, 569–585. [CrossRef]
2. Paprotny, D.; Sebastian, A.; Morales-Nápoles, O.; Jonkman, S. Trends in flood losses in Europe over the past 150 years. *Nat. Commun.* **2018**, *9*, 1985. [CrossRef] [PubMed]
3. Ávila, Á.; Guerrero, F.C.; Escobar, Y.C.; Justino, F. Recent Precipitation Trends and Floods in the Colombian Andes. *Water* **2019**, *11*, 379. [CrossRef]
4. Mirza, M.M.Q. Climate change, flooding in South Asia and implications. *Reg. Environ. Chang.* **2011**, *11*, 95–107. [CrossRef]
5. Sofia, G.; Roder, G.; Fontana, G.D.; Tarolli, P. Flood dynamics in urbanised landscapes: 100 years of climate and humans' interaction. *Sci. Rep.* **2017**, *7*, 40527. [CrossRef] [PubMed]
6. Min, S.-K.; Zhang, X.; Zwiers, F.W.; Hegerl, G.C. Human contribution to more-intense precipitation extremes. *Nature* **2011**, *470*, 378–381. [CrossRef] [PubMed]
7. Chang, L.-C.; Chang, F.-J.; Yang, S.-N.; Kao, I.-F.; Ku, Y.-Y.; Kuo, C.-L.; Amin, I.R.; bin Mat, M.Z. Building an Intelligent Hydroinformatics Integration Platform for Regional Flood Inundation Warning Systems. *Water* **2019**, *11*, 9. [CrossRef]
8. Célleri, R.; Feyen, J. The Hydrology of Tropical Andean Ecosystems: Importance, Knowledge Status, and Perspectives. *Mt. Res. Dev.* **2009**, *29*, 350–355. [CrossRef]
9. Muñoz, P.; Célleri, R.; Feyen, J. Effect of the Resolution of Tipping-Bucket Rain Gauge and Calculation Method on Rainfall Intensities in an Andean Mountain Gradient. *Water* **2016**, *8*, 534. [CrossRef]
10. Arias, P.A.; Garreaud, R.; Poveda, G.; Espinoza, J.C.; Molina-Carpio, J.; Masiokas, M.; Viale, M.; Scaff, L.; van Oevelen, P.J. Hydroclimate of the Andes Part II: Hydroclimate Variability and Sub-Continental Patterns. *Front. Earth Sci.* **2021**, *8*, 666. [CrossRef]
11. Hundecha, Y.; Parajka, J.; Viglione, A. Flood type classification and assessment of their past changes across Europe. *Hydrol. Earth Syst. Sci. Discuss.* **2017**, 1–29.
12. Turkington, T.; Breinl, K.; Ettema, J.; Alkema, D.; Jetten, V. A new flood type classification method for use in climate change impact studies. *Weather. Clim. Extrem.* **2016**, *14*, 1–16. [CrossRef]
13. Borga, M.; Gaume, E.; Creutin, J.D.; Marchi, L. Surveying flash floods: Gauging the ungauged extremes. *Hydrol. Process.* **2008**, *22*, 3883–3885. [CrossRef]
14. Knocke, E.T.; Kolivras, K.N. Flash Flood Awareness in Southwest Virginia. *Risk Anal. Int. J.* **2007**, *27*, 155–169. [CrossRef] [PubMed]
15. Sottolichio, A.; Hurther, D.; Gratiot, N.; Bretel, P. Acoustic turbulence measurements of near-bed suspended sediment dynamics in highly turbid waters of a macrotidal estuary. *Cont. Shelf Res.* **2011**, *31*, S36–S49. [CrossRef]
16. Borga, M.; Anagnostou, E.; Blöschl, G.; Creutin, J.-D. Flash flood forecasting, warning and risk management: The HYDRATE project. *Environ. Sci. Policy* **2011**, *14*, 834–844. [CrossRef]
17. del Granado, S.; Stewart, A.; Borbor, M.; Franco, C.; Tauzer, E.; Romero, M. *Flood Early Warning Systems. Comparative Analysis in Three Andean Countries (Sistemas de Alerta Temprana para Inundaciones: Análisis Comparativo de Tres Países Latinoamericanos)*; Institute for Advanced Development Studies (INESAD): La Paz, Bolivia, 2016. (In Spanish)
18. Noymanee, J.; Theeramunkong, T. Flood forecasting with machine learning technique on hydrological modeling. *Procedia Comput. Sci.* **2019**, *156*, 377–386. [CrossRef]
19. Dávila, D. *Flood Warning Systems in Latin America (21 Experiencias de Sistemas de Alerta Temprana en América Latina)*; Soluciones Prácticas: Lima, Peru, 2016.
20. Boers, N.; Bookhagen, B.; Barbosa, H.D.M.J.; Marwan, N.; Kurths, J.; Marengo, J.A. Prediction of extreme floods in the eastern Central Andes based on a complex networks approach. *Nat. Commun.* **2014**, *5*, 5199. [CrossRef]
21. Aybar Camacho, C.L.; Lavado-Casimiro, W.; Huerta, A.; Fernández Palomino, C.; Vega-Jácome, F.; Sabino Rojas, E.; Felipe-Obando, O. Use of the gridded product 'PISCO' for precipitation studies, investigations and operationl systems of monitoring and hydrometeorological forecasting (Uso del Producto Grillado 'PISCO' de precipitación en Estudios, Investigaciones y Sistemas Operacionales de Monitoreo y Pronóstico Hidrometeorológico). *Nota Técnica* **2017**. No. 001 SENAMHI-DHI.
22. Fernández de Córdova Webster, C.J.; Rodríguez López, Y. First results of the current hydrometeorological network of Cuenca, Ecuador(Primeros resultados de la red actual de monitoreo hidrometeorológico de Cuenca, Ecuador). *Ing. Hidráulica Ambient.* **2016**, *37*, 44–56.
23. Clark, M.P.; Bierkens, M.F.P.; Samaniego, L.; Woods, R.A.; Uijlenhoet, R.; Bennett, K.E.; Pauwels, V.R.N.; Cai, X.; Wood, A.W.; Peters-Lidard, C.D. The evolution of process-based hydrologic models: Historical challenges and the collective quest for physical realism. *Hydrol. Earth Syst. Sci.* **2017**, *21*, 3427–3440. [CrossRef]

24. Mosavi, A.; Ozturk, P.; Chau, K.-W. Flood Prediction Using Machine Learning Models: Literature Review. *Water* **2018**, *10*, 1536. [CrossRef]
25. Young, P.C. Advances in real-time flood forecasting. *Philos. Trans. R. Soc. Lond. Ser. A Math. Phys. Eng. Sci.* **2002**, *360*, 1433–1450. [CrossRef] [PubMed]
26. Bontempi, G.; Ben Taieb, S.; Le Borgne, Y.-A. Machine Learning Strategies for Time Series Forecasting. In *Lecture Notes in Business Information Processing*; Springer: Berlin/Heidelberg, Germany, 2013; pp. 62–77.
27. Galelli, S.; Castelletti, A. Assessing the predictive capability of randomized tree-based ensembles in streamflow modelling. *Hydrol. Earth Syst. Sci.* **2013**, *17*, 2669–2684. [CrossRef]
28. Muñoz, P.; Orellana-Alvear, J.; Willems, P.; Célleri, R. Flash-Flood Forecasting in an Andean Mountain Catchment—Development of a Step-Wise Methodology Based on the Random Forest Algorithm. *Water* **2018**, *10*, 1519. [CrossRef]
29. Furquim, G.; Neto, F.; Pessin, G.; Ueyama, J.; Joao, P.; Clara, M.; Mendiondo, E.M.; de Souza, V.C.; de Souza, P.; Dimitrova, D.; et al. Combining wireless sensor networks and machine learning for flash flood nowcasting. In Proceedings of the 2014 28th International Conference on Advanced Information Networking and Applications Workshops, Victoria, BC, Canada, 13–16 May 2014; pp. 67–72.
30. Toukourou, M.; Johannet, A.; Dreyfus, G.; Ayral, P.-A. Rainfall-runoff modeling of flash floods in the absence of rainfall forecasts: The case of 'Cévenol flash floods. *Appl. Intell.* **2011**, *35*, 178–189. [CrossRef]
31. Adamowski, J.F. Development of a short-term river flood forecasting method for snowmelt driven floods based on wavelet and cross-wavelet analysis. *J. Hydrol.* **2008**, *353*, 247–266. [CrossRef]
32. Aichouri, I.; Hani, A.; Bougherira, N.; Djabri, L.; Chaffai, H.; Lallahem, S. River Flow Model Using Artificial Neural Networks. *Energy Procedia* **2015**, *74*, 1007–1014. [CrossRef]
33. Khosravi, K.; Shahabi, H.; Pham, B.T.; Adamowski, J.; Shirzadi, A.; Pradhan, B.; Dou, J.; Ly, H.-B.; Gróf, G.; Ho, H.L.; et al. A comparative assessment of flood susceptibility modeling using Multi-Criteria Decision-Making Analysis and Machine Learning Methods. *J. Hydrol.* **2019**, *573*, 311–323. [CrossRef]
34. Solomatine, D.P.; Xue, Y. M5 Model Trees and Neural Networks: Application to Flood Forecasting in the Upper Reach of the Huai River in China. *J. Hydrol. Eng.* **2004**, *9*, 491–501. [CrossRef]
35. Khalaf, M.; Hussain, A.J.; Al-Jumeily, D.; Fergus, P.; Idowu, I.O. Advance flood detection and notification system based on sensor technology and machine learning algorithm. In Proceedings of the 2015 International Conference on Systems, Signals and Image Processing (IWSSIP), London, UK, 10–12 September 2015; pp. 105–108.
36. Contreras, P.; Orellana-Alvear, J.; Muñoz, P.; Bendix, J.; Célleri, R. Influence of Random Forest Hyperparameterization on Short-Term Runoff Forecasting in an Andean Mountain Catchment. *Atmosphere* **2021**, *12*, 238. [CrossRef]
37. Orellana-Alvear, J.; Célleri, R.; Rollenbeck, R.; Muñoz, P.; Contreras, P.; Bendix, J. Assessment of Native Radar Reflectivity and Radar Rainfall Estimates for Discharge Forecasting in Mountain Catchments with a Random Forest Model. *Remote Sens.* **2020**, *12*, 1986. [CrossRef]
38. Razali, N.; Ismail, S.; Mustapha, A. Machine learning approach for flood risks prediction. *IAES Int. J. Artif Intell.* **2020**, *9*, 73. [CrossRef]
39. Chen, S.; Xue, Z.; Li, M. Variable Sets principle and method for flood classification. *Sci. China Ser. E Technol. Sci.* **2013**, *56*, 2343–2348. [CrossRef]
40. Jati, M.I.H.; Santoso, P.B. Prediction of flood areas using the logistic regression method (case study of the provinces Banten, DKI Jakarta, and West Java). *J. Phys. Conf. Ser.* **2019**, *1367*, 12087. [CrossRef]
41. Bishop, C.M. *Pattern Recognition and Machine Learning*; Springer: New York, NY, USA, 2006.
42. Dodangeh, E.; Choubin, B.; Eigdir, A.N.; Nabipour, N.; Panahi, M.; Shamshirband, S.; Mosavi, A. Integrated machine learning methods with resampling algorithms for flood susceptibility prediction. *Sci. Total. Environ.* **2020**, *705*, 135983. [CrossRef] [PubMed]
43. Breiman, L. Random forests. *Mach. Learn.* **2001**, *45*, 5–32. [CrossRef]
44. Breiman, L. *Classification and Regression Trees*; Routledge: Oxfordshire, UK, 2017.
45. Zadrozny, B.; Elkan, C. Obtaining calibrated probability estimates from decision trees and naive bayesian classifiers. In Proceedings of the Eighteenth International Conference on Machine Learning, Williamstown, MA, USA, 28 June–1 July 2001; Morgan Kaufmann Publishers, Inc.: San Francisco, CA, USA; pp. 609–616.
46. Zhang, H. The optimality of naive Bayes. *AA* **2004**, *1*, 3.
47. Maier, H.R.; Jain, A.; Dandy, G.C.; Sudheer, K. Methods used for the development of neural networks for the prediction of water resource variables in river systems: Current status and future directions. *Environ. Model. Softw.* **2010**, *25*, 891–909. [CrossRef]
48. Maier, H.R.; Dandy, G.C. Neural networks for the prediction and forecasting of water resources variables: A review of modelling issues and applications. *Environ. Model. Softw.* **2000**, *15*, 101–124. [CrossRef]
49. Sudheer, K.P.; Gosain, A.K.; Ramasastri, K.S. A data-driven algorithm for constructing artificial neural network rainfall-runoff models. *Hydrol. Process.* **2002**, *16*, 1325–1330. [CrossRef]
50. Pedregosa, F.; Varoquaux, G.; Gramfort, A.; Michel, V.; Thirion, B.; Grisel, O.; Blondel, M.; Prettenhofer, P.; Weiss, R.; Dubourg, V.; et al. Scikit-learn: Machine Learning in {P}ython. *J. Mach. Learn. Res.* **2011**, *12*, 2825–2830.
51. Chen, C.; Liaw, A.; Breiman, L. *Using Random Forest to Learn Imbalanced Data*; University of California: Berkeley, CA, USA, 2004; Volume 110, pp. 1–12.
52. Read, J.; Pfahringer, B.; Holmes, G.; Frank, E. Classifier chains for multi-label classification. *Mach. Learn.* **2011**, *85*, 333–359. [CrossRef]

53. Gu, Q.; Zhu, L.; Cai, Z. Evaluation Measures of the Classification Performance of Imbalanced Data Sets. In *Computational Intelligence and Intelligent Systems*; Springer: Berlin/Heidelberg, Germany, 2009; Volume 51, pp. 461–471. [CrossRef]
54. Sun, Y.; Wong, A.K.C.; Kamel, M.S. Classification of imbalanced data: A review. *Int. J. Pattern Recognit. Artif. Intell.* **2009**, *23*, 687–719. [CrossRef]
55. Hossin, M.; Sulaiman, M.N. A review on evaluation metrics for data classification evaluations. *Int. J. Data Min. Knowl. Manag. Process.* **2015**, *5*, 1.
56. Akosa, J. Predictive accuracy: A misleading performance measure for highly imbalanced data. In Proceedings of the SAS Global Forum, Orlando, FL, USA, 2–5 April 2017; pp. 2–5.
57. Raschka, S. Model evaluation, model selection, and algorithm selection in machine learning. *arXiv* **2018**, arXiv:1811.12808.
58. de Almeida, I.K.; Almeida, A.K.; Anache, J.A.A.; Steffen, J.L.; Sobrinho, T.A. Estimation on time of concentration of overland flow in watersheds: A review. *Geociencias* **2014**, *33*, 661–671.
59. Loumagne, C.; Normand, M.; Riffard, M.; Weisse, A.; Quesney, A.; Le Hégarat-Mascle, S.; Alem, F. Integration of remote sensing data into hydrological models for reservoir management. *Hydrol. Sci. J.* **2001**, *46*, 89–102. [CrossRef]
60. Li, Y.; Grimaldi, S.; Walker, J.P.; Pauwels, V.R.N. Application of Remote Sensing Data to Constrain Operational Rainfall-Driven Flood Forecasting: A Review. *Remote Sens.* **2016**, *8*, 456. [CrossRef]

 *hydrology* 

Article

Real-Time Flood Mapping on Client-Side Web Systems Using HAND Model

Anson Hu [1] **and Ibrahim Demir** [2,*]

[1] Computer Science, Massachusetts Institute of Technology, Cambridge, MA 02139, USA; ansonhu@mit.edu
[2] Civil and Environmental Engineering, University of Iowa, Iowa City, IA 52246, USA
* Correspondence: ibrahim-demir@uiowa.edu

Abstract: The height above nearest drainage (HAND) model is frequently used to calculate properties of the soil and predict flood inundation extents. HAND is extremely useful due to its lack of reliance on prior data, as only the digital elevation model (DEM) is needed. It is close to optimal, running in linear or linearithmic time in the number of cells depending on the values of the heights. It can predict watersheds and flood extent to a high degree of accuracy. We applied a client-side HAND model on the web to determine extent of flood inundation in several flood prone areas in Iowa, including the city of Cedar Rapids and Ames. We demonstrated that the HAND model was able to achieve inundation maps comparable to advanced hydrodynamic models (i.e., Federal Emergency Management Agency approved flood insurance rate maps) in Iowa, and would be helpful in the absence of detailed hydrological data. The HAND model is applicable in situations where a combination of accuracy and short runtime are needed, for example, in interactive flood mapping and supporting mitigation decisions, where users can add features to the landscape and see the predicted inundation.

Keywords: flood maps; flood risk management; HAND model; WebAssembly; flood risk mapping; web systems; floods; urban flooding; flood analysis

Citation: Hu, A.; Demir, I. Real-Time Flood Mapping on Client-Side Web Systems Using HAND Model. *Hydrology* **2021**, *8*, 65. https://doi.org/10.3390/hydrology8020065

Academic Editors: Evangelos Baltas, Andrea Petroselli and Marina Iosub

Received: 26 January 2021
Accepted: 3 April 2021
Published: 11 April 2021

1. Introduction

Flooding is a global problem, and floods around the world are becoming more significant and severe [1]. Of natural disasters worldwide 40% are floods [2]. It is important to model and predict inundation extent of floods, which is critical information for flood mitigation [3,4], preparedness [5], and planning and response efforts [6]. Real-time and accurate predictions of flood inundation extent can facilitate understanding the potential flood risk and damage [7,8], and support flood mitigation and planning [9]. Flood inundation maps can be used in flood risk communication using intelligent systems [10,11] and novel communication systems [12].

Floods are caused by a multitude of factors, which makes prediction difficult [13]. Moisture flows and consequently rainfall [14], runoff, and stream behavior [15] and other issues affect flooding significantly. For these reasons, flood prediction has a highly complex and non-linear nature [16]. There are various models to predict flood extent. While empirical models include observations [17] and remote sensing [18], and benchmark datasets [19] of past events, hydrodynamic models simulate physical movement of the water [20]. Data driven methods are used extensively in hydrology and water resources modeling [21,22]. One of the simplest and earliest data driven methods based on elevation is flood-fill, which produces results in optimal time but provides less accurate predictions. Another simple method involves overlaying an observed flood level with the topography [23]. On the other hand, there are a variety of complex hydrological models that take into account numerous distinct factors, such as the three-dimensional model [24]. Generally, models that have a good balance between speed and accuracy are used in operational needs [25].

There are several challenges with the existing hydrodynamic models to be helpful in real-time and operational decision-making situations [26]. One of the main issues is that these models often require a lot of detailed hydrological data [27]. For example, advanced hydrodynamic models for flood map prediction often take into account a plethora of factors, including riverbed dynamics, hydraulic features, and structures [28]. The second main challenge to utilize these models in decision making is their computational complexity. Some of the methods that provide a balance between speed and accuracy [29] also utilize data that is costly to gather, such as topographical indices [30].

The height above the nearest drainage algorithm has a wide variety of applications. The height above nearest drainage (HAND) method was originally designed to categorize soil environments and find water tables. Renno et al. [31] first outlined the algorithm and applied it to categorize swampy forests vs. terra-firme forests in the Amazonian jungle. Nobre et al. [32] applied HAND to categorize water tables in a large area of the central Amazon rainforest. The most recent application area for HAND methods is in flood inundation. Nobre et al. [33] applied the HAND model to predict flooding in Southern Brazil. The largest application of this is being able to predict flooding in areas that have little to no prior data, due to the HAND model's total reliance on the DEM. Previous studies on HAND model to predict flood maps [33,34] achieved model accuracy around 86-98% when compared to reference flood extent maps.

Recent advancements in web technologies have allowed for desktop level computation and analysis of large-scale datasets. Web systems have been effectively used in environmental data analysis [35,36], distributed computing [37], and geospatial data processing [38]. WebAssembly is a low-level assembly-like language for the web with near-native performance and allows languages such as C/C++, C# and Rust to run on the web. WebAssembly was recently released in March 2017 and is designed to run code both safely and efficiently in browsers on the client-side [39]. These technologies have allowed for previously unobtainable speeds and computations to be performed on local users' computers.

Here, we utilize the HAND method in order to provide flood inundation extent predictions in real-time on client-side web systems. We discuss the applicability of the algorithm in comparison with other prediction algorithms for flood-prone regions of Iowa and the applications for such real-time prediction. Our objective is to outline the HAND algorithm and demonstrate that it has potential to predict both accurately and quickly on the web, and support flood related decision-making activities [40,41]. There are a variety of benefits that HAND can provide in flood inundation prediction. First, due to its low complexity, HAND runs in very fast time both theoretically and practically. This means that stakeholders involved or interested in flood planning (for example a government official or concerned citizen) can easily make changes to a DEM to simulate adding man-made barriers and then run the algorithm to determine how these changes affect flood inundation extents. This can lead to more effective flood protection planning, since previously users were meant to look at a standalone flood model generated once by an advanced but computationally expensive algorithm and then make educated guesses on the design and utilization of man-made flood prevention measures.

In the following sections, we will outline the algorithm, details of implementations on the web systems, and several evaluation case-studies for comparison to real-world flooding scenarios.

2. Materials and Methods

Our goal with this study is to demonstrate the implementation of the HAND model on client-side web systems and evaluate the accuracy and performance for real-time decision-making scenarios. The system will provide an implementation of the HAND algorithm that runs in the local browser of the user for fast flood extent prediction that can work offline and in operational settings with minimal processing on the web server. The system will allow users to modify aspects of the terrain for mitigation purposes and be able to see predicted flood inundation extents in order to facilitate the use of measures to control potential flood damage.

2.1. Web Implementation of HAND

Figure 1 presents the architecture of the real-time web-based HAND flood map generation system. It includes several steps to compute and generate flood maps using the HAND algorithm. The system is designed so that the algorithmic portion of HAND is entirely in C++ and run on WebAssembly to enable client-side implementation. The code for the computationally heavy portion of the HAND algorithm was written in C++. Emscripten was used to compile the C++ code to WebAssembly, which allows JavaScript (JS) to pass input data and receive model output. The only input is the DEM data and the drainage threshold value provided by the user. The WebAssembly-compiled implementation of HAND calculates the heights above nearest drainages and passes that information to JavaScript for visualizing the flood extent on Google Maps.

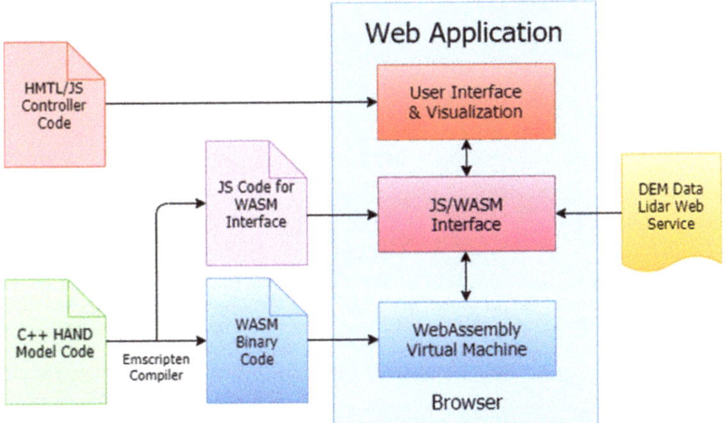

Figure 1. Architecture and components of the real-time web-based height above nearest drainage (HAND) mapping system.

The C++ algorithmic implementation is precompiled into WebAssembly. The resulting code is meant to supply an efficient way to run the HAND algorithm. Most features of prediction, which can be applied either pre- or post-computation, are left out of the WebAssembly component. WebAssembly is running on an in-browser virtual machine. JavaScript glue code generated by Emscripten provides a means to interface with the WebAssembly VM. The HTML/JS frontend loads images from a source, either static or online, and applies user selected parameters before and after computation. This piece of the architecture calls the API provided by the glue code to interface with the WebAssembly VM and provide prediction results. The JS manages the frontend of the application for taking user inputs, integrating external terrain data, and visualizing map output.

The web implementation is meant to make the system as flexible as possible for all sorts of applications that can utilize the HAND algorithm implementation, and apply their own features (changing terrain, allowing for multiple parameter changes in local areas, etc.). For example, one application might scrape DEM data from various online sources and allow users to generate flood maps in any region in the world. A different application might generate DEMs with another algorithm and allow users to explore the generated flood maps.

Many of the features relevant for flood prediction are nearly independent of the computation of the HAND. For example, two of the three relevant parameters, HAND threshold and absolute height threshold, can be applied after the actual HAND model is computed. In addition, other features such as applying HAND thresholds only to specific areas, dynamically loading DEMs, and adding terrain features (dikes, reservoirs, etc.) are better suited to the client-side processing using JavaScript. Since these features are highly application-specific but low in terms of computational expense, we leave them out of the C++ and WebAssembly implementation. For example, one application might require dynamic loading of DEMs from state resources, while another generates DEMs with an algorithm and passes them in. This allows the C++ implementation to be highly portable and versatile; a wide variety of applications requiring the use of the HAND algorithm can utilize it.

2.2. *HAND Algorithm*

In our study, we applied HAND methodology (as described in [32,33]) in order to predict the extent of flood inundation. By calculating the vertical distance from a cell and its parent cell and applying several heuristic steps, we can calculate the extent of flood inundation with a comparable accuracy. A brief description of the steps (Figure 2) for HAND algorithm is given here:

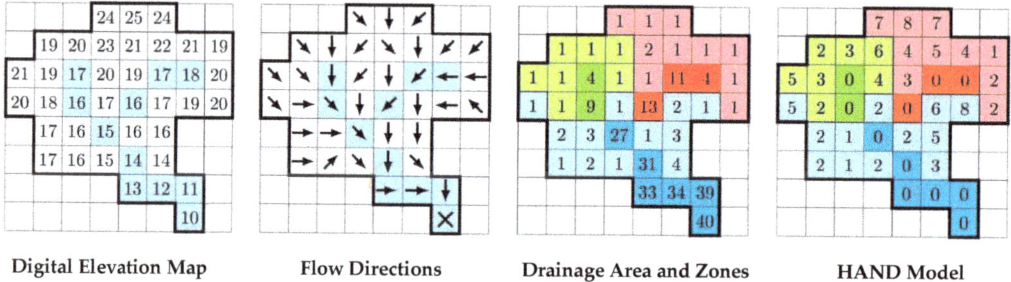

Figure 2. Steps of HAND algorithm procedure to generate flood maps (adapted from [42]).

Determining Flow Direction: We define the direction of the flow of cell a in the DEM as the adjacent cell to which the water on a will flow. Water can only flow in one direction, which means that it can only flow to one cell. We define the set of adjacent cells to the cell at coordinates (x0, y0) in Equation (1) as

$$\{ (x, y) \mid \max (|x - x0|, |y - y0|) = 1 \} \tag{1}$$

We define the child of cell a to be cell b such that the direction of the flow of cell a is towards cell b. We define a flow chain of cell a as an ordered list of cells such that the first cell in the list is a, and each subsequent cell is a child of the previous cell.

Depression Removal: First, the graph must be normalized to remove depressions that are undrainable. An undrainable depression is a set of cells in which any pair of cells in the set can be reached by traversing a path of cells in which consecutive cells on the path are adjacent to each other and each cell in the path is part of the set and none of the cells have a continuous flow to the border of the DEM (none of the cells have a border cell in their flow chain). We used priority-flood depression filling [43] in this case. This ensures that all water can drain to the outside of the map, and that there will be no cycles or other anomalies that can disrupt subsequent algorithms. This step runs in O(N) for integer elevations or O(N log N) for floating-point elevations, where N is the number of cells.

Computing Flow Directions: Then, the direction of water flow for each cell needs to be resolved. We used the D8 algorithm to compute the flow directions. The D8 algorithm takes the adjacent cell that provides the highest gradient to be the cell of the direction of the flow. A flat-resolving algorithm [44] was used to determine the direction of flow in flat areas (where all adjacent cells to a cell have the same height) and guide water away from high areas and towards lower areas. This step runs in O(N) time.

HAND Model Generation: Next, drainages need to be established on the graph. This is done by setting an accumulated area threshold. If a cell has more units of area of water that drain through it than the area threshold, then it becomes a drainage cell. In other words, if the number of flow chains that the cell is part of is greater than the accumulated area threshold, then it is a drainage cell. The nearest drainage for a cell is then the first cell in its flow chain that is a drainage cell. This step runs in O(N) time. Finally, the heights above the nearest drainage are calculated by subtracting the height of the cell in the original DEM with the nearest drainage cell's height in the original DEM.

2.3. Implementation Challenges

There were several issues encountered during the course of implementing the HAND model on the web. HAND model does not account for some of the parameters like height of water and absolute height.

Height of Water: The HAND algorithm as it currently exists views each drainage stream as only one pixel in width, and it does not take into account the height of water. This means that the algorithm will think that a wide river is actually a canyon. The algorithm thinks that the banks of the river have extremely high HAND values because the stream's elevation is at the bottom of the river. In reality the banks of the river have HAND value near zero, because water goes up to the top of the river bank, but the algorithm does not reflect this. To overcome this issue, we added two parameters into implementation.

(a) HAND flood threshold (hereafter referred to as "HAND threshold"). This is the main component of the flood inundation calculation. If the HAND of a cell is below this threshold, then it will be counted as flooded, as long as it satisfies the third parameter. In other words, cell a is counted as flooded by threshold t if HAND[a] < t.
(b) Accumulated area of drainage threshold (hereafter referred to "drainage threshold"). This is used to define which flows of water are actually drainages. If the amount of water area that flows over a cell is over this threshold, then it is counted as drainage.

There are several effects of these parameters on the model outcome. As the HAND threshold goes up, the extent of areas marked as flooded will increase. If the HAND threshold is defined too high, the entire area will be flooded. Drainage threshold controls how many drainage points are available in the selected region. If the drainage threshold is defined too high, the algorithm becomes a slow flood-fill. If it is too low, the flooding predictions are exceedingly coarse, with big blotches splattered around with general inaccuracy. By having a high HAND threshold and enough drainage streams, we allow the areas adjacent to the rivers to be easily flooded, despite being viewed as canyons by the algorithm.

Absolute Height: If there is a large plateau that drains into a stream going downhill, the algorithm will view the plateau as flooded, even when it might be far above local rivers and other water bodies and have a low chance of flooding. To overcome this issue, we added a third parameter for the maximum height threshold of flooding (hereafter referred to "absolute height threshold"). If a cell's absolute height is above this threshold, then it will not be considered as flooded. Absolute height threshold controls extent of flooding. If it is defined too high, areas that would be dry are predicted as flooded. If it is too low, the algorithm will miss much of the flooding in higher areas. A medium absolute height threshold allows rivers that have a slope to still have effective flooding predictions within them and on their borders but prevents adjacent flat-but-elevated landforms being counted as flooded. A combination of a medium absolute height threshold and high HAND threshold effectively serves to counter the two issues listed.

3. Results and Discussion

We compared the results of the HAND output with two other flood inundation predictors: a basic flood-fill implementation and Federal Emergency Management Agency (FEMA) approved Flood Insurance Rate Maps (100-yr, 500-yr) developed [45] at the Iowa Flood Center (IFC). We used two datasets for the analysis. The first was the Cedar Rapids (Figure 3) metro area in Iowa, where we utilized a DEM at a resolution of 1 m for 5 sq km area (2863 × 1746 grid). The second was for the Ames, Iowa (Figure 3), where we used a DEM at a resolution of 2 m for 53.8 sq km area (4220 × 3189 grid). All DEMs were normalized to 255 units corresponding to 1–3 ft vertical elevation in real-world based on vertical resolution.

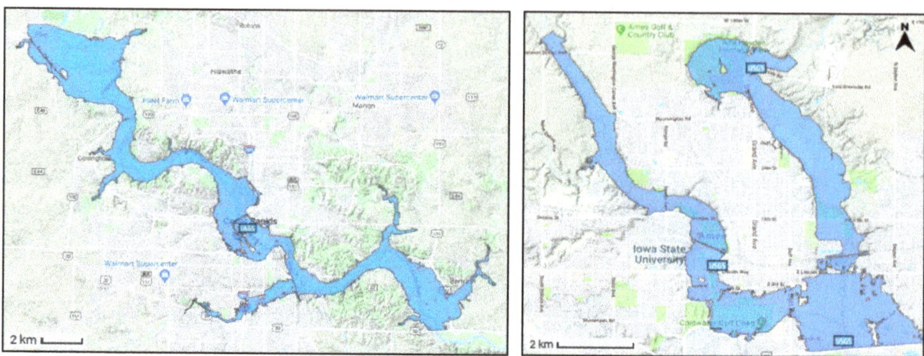

Figure 3. FEMA-approved flood maps for Cedar Rapids (**left**) and Ames (**right**), Iowa by the Iowa Flood Center (IFC).

For the Cedar Rapids area, the HAND algorithm ran in roughly 3 s using a mid-range personal laptop and produced an output differing from the reference map predictions by 10%. The flood-fill implementation ran in less than one second and produced an output differing by roughly 15%. For the Ames area, the HAND algorithm (Figure 4) ran in roughly 9 s and produced an output differing from the reference map predictions by 6%. The flood-fill implementation (Figure 4) ran in roughly one second and produced an output differing by roughly 30%. The HAND model's accuracy was notably improved, both in absolute error and in error relative to the flood-fill implementation; this was most likely due to the larger scale of the Ames DEM. Comparisons were drawn between appropriate thresholds in the FEMA reference maps and the results of the HAND implementation. Several predictions of arbitrary values of parameters were generated with the algorithm and are presented here to show the effects of adjusting the parameters of the prediction algorithm on the output.

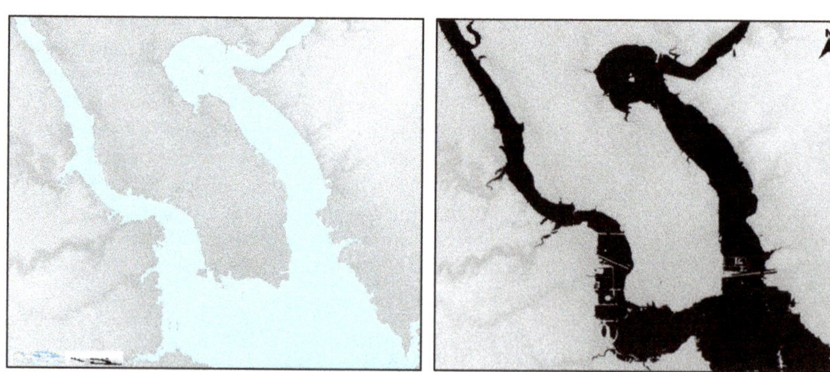

Figure 4. Maps generated for Ames, Iowa using basic flood-fill (**left**) and HAND model (**right**).

3.1. Adjustment of Parameters

To adjust for flood predictions in areas without prior predictions or flood maps, we must take into account the scale of the area we are modifying, and a sense of an absolute maximum for good area control. The Ames area DEM measures 65 square km, and the most effective drainage threshold measured was 1 million cells corresponding to a 4 square km area. This is most likely the absolute maximum, as it is strikingly similar to suggested effective threshold value of 4.05 square km in the literature [33]. The HAND threshold is highly correlated with the level of the flooding on the terrain. The best way to analyze this is to figure out the relative heights of the cells and compare them to real-world heights, and then scale appropriately. For example, if the cells had relative heights corresponding to one unit as 2 ft, and if information exists that areas less than 4 ft above the nearest drainage will be flooded, then the HAND threshold can be set as two units corresponding to 4 ft.

The absolute threshold must be determined by the maximum height that the user predicts the flood will extend, similar to the HAND threshold, but in an absolute manner. To generalize parameters to areas for which there is no prior prediction available, it is recommended that areas with similar topography and known predictions for flood inundation extents be examined; then, the user can find good parameters for the HAND algorithm that match those known predictions, and apply those same parameters to the unknown area.

3.2. Evaluation of Parameter Sensitivity

We evaluated the sensitivity of the HAND parameters on generating the flood extent. Tables 1 and 2 present the resulting flood extents for changing model parameters mentioned in the HAND algorithm section above. We tested several drainage threshold (100 K, 500 K, 1 M, and 2 M cells), HAND threshold (2, 3, 4, and 6 units), and absolute elevation threshold (220 and 225 ft) values. As seen from the resulting flood extents, higher HAND threshold values cause larger areas around drainage points to be flooded. Higher drainage threshold on the other hand causes number of drainage points to be fewer, which causes less branching in the predictions. Higher absolute threshold causes fewer high-elevation areas to be flooded. These parameters have significant impact on the resulting map details and accuracy. Drainage threshold has an optimal value as 4 square kilometers as supported by the literature [33], however it can further be calibrated if reference maps are available for selected region. The HAND threshold represents the water level at the outlet of the selected watershed. The HAND threshold value can be assigned using visual inspection, water level gauge, forecast models or historical data from a near-by gauge.

Table 1. Sensitivity of model parameters (HAND threshold, drainage threshold) on the resulting flood extent (absolute threshold 220 ft).

	HAND Threshold			
Drainage Threshold	**2**	**3**	**4**	**6**
100 k cells	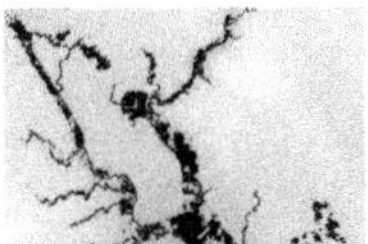			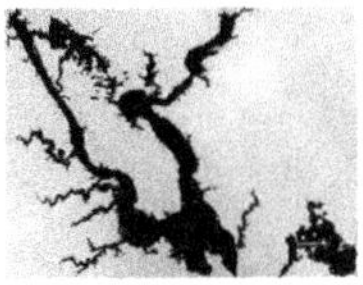
500 k cells	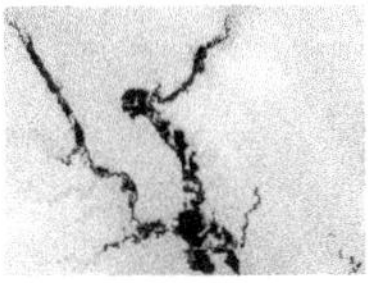	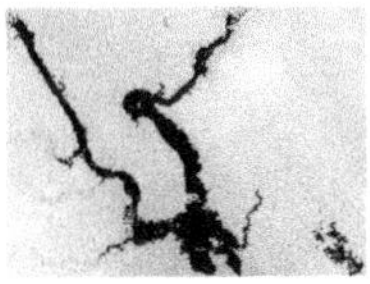		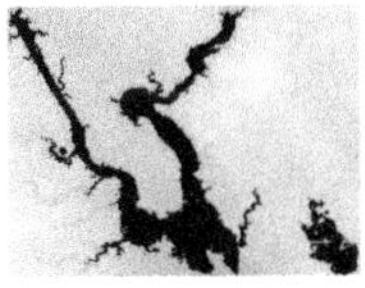

Table 1. *Cont.*

	HAND Threshold			
Drainage Threshold	**2**	**3**	**4**	**6**
1 m cells				
2 m cells				

Table 2. Sensitivity of model parameters (HAND threshold and drainage threshold) on the resulting flood extent (absolute threshold 225 ft).

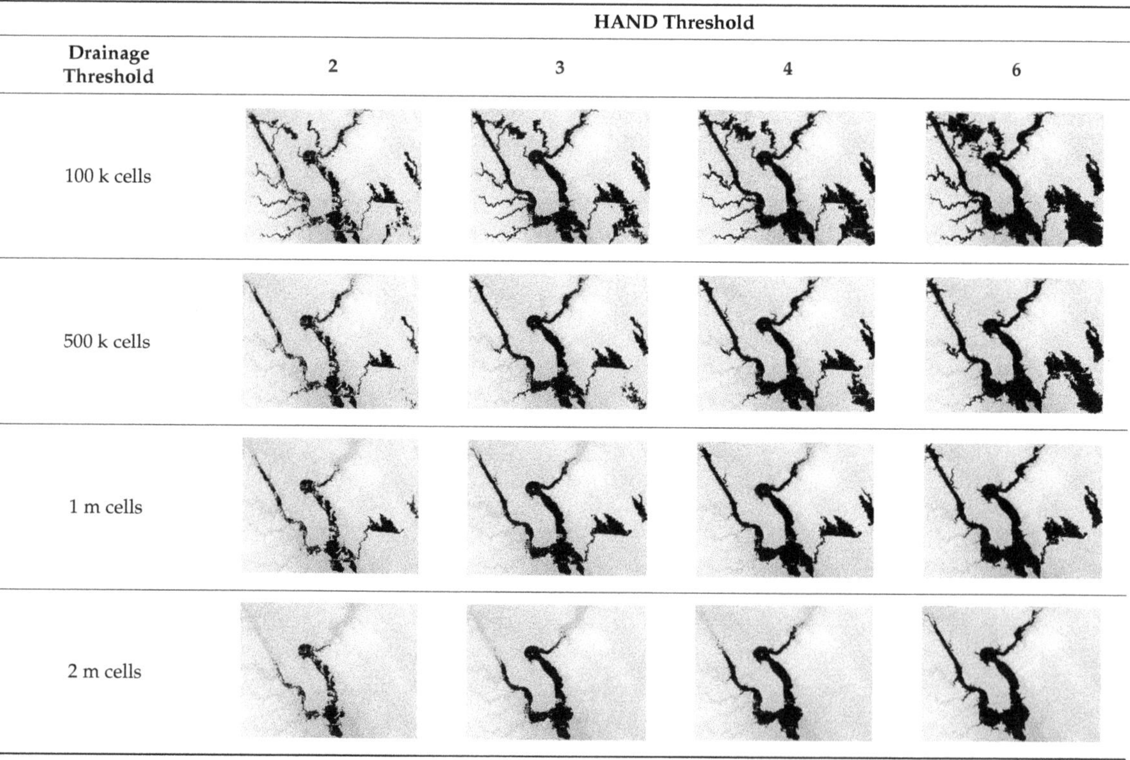

	HAND Threshold			
Drainage Threshold	2	3	4	6
100 k cells				
500 k cells				
1 m cells				
2 m cells				

3.3. Real-Time HAND Generation Performance

In this study, we evaluated the computational performance of client-side flood map generation using the HAND model and WebAssembly. We selected different DEM resolutions (1, 5, 10, 25, 50, and 100 m) and grid sizes (1, 4, 9, 16, and 25 million cells) for evaluation. Based on the selected elevation data resolution and grid size, generated maps represent city, county, and state scale in coverage (Table 3). The performance of the map generation took roughly 0.25–6.25 s on a moderate personal computer. Higher DEM resolutions could be used for any scale based on data availability and computational resources. The client-side implementation on WebAssembly could further speed up by using Web Workers and distributing the computation to multiple CPUs on the same machine.

Table 3. Computing time for generating flood extent in different scales and resolutions.

Scale	DEM Resolution	Region	Grid Cells	Computing Time
City	1–5 m	1–225 sq km	1–9 million	0.25–2.25 s
County	5–10 m	225–1600 sq km	9–16 million	2.25–4.00 s
State	25–100 m	10,000–250,000 sq km	16–25 million	4.00–6.25 s

Terrain Modification and Mitigation Scenarios: The computational speed and starting the mapping process from bare elevation dataset is a critical benefit of the presented system for what if scenario analysis in flood mitigation. The users can modify the terrain for potential flood preparedness and mitigation applications (i.e., adding flood walls, elevating structures, and sand bagging critical infrastructure) and generate the resulting map within seconds to evaluate the mitigation application. The system can use base or modified terrain data as an input and generate corresponding floods maps. Users can evaluate mitigation options based on resulting flood maps and how much area is saved from flooding.

A sample case study was generated for Cedar Rapids, Iowa, where resulting map can be seen in Figure 5 before and after terrain modification. Mitigation impact of the levee on resulting flood can be seen from the figure. Significant area in the urban setting could be saved from flooding by installing a levee on the west site of the river. The system can be used for what-if scenario evaluation by simple terrain modifications. While this analysis could take hours to days to run on traditional flood mapping applications, it takes under a minute to modify the terrain and generate the resulting maps in the proposed system.

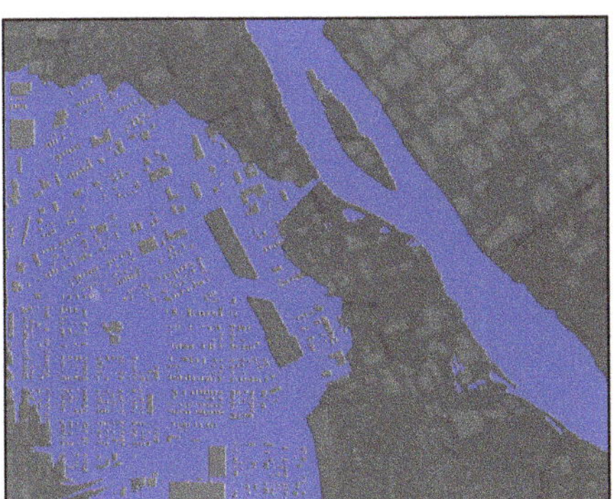
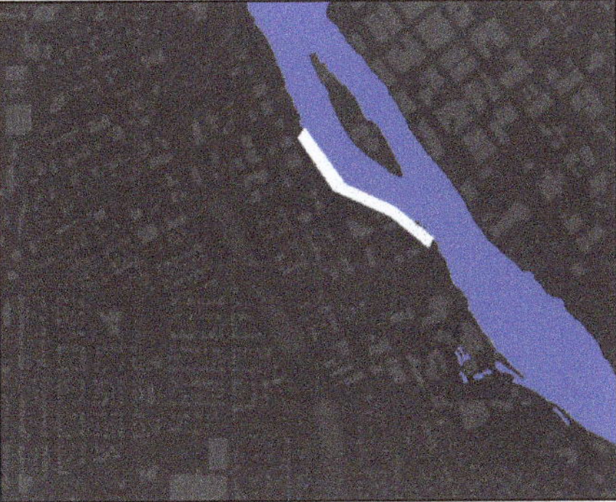

Figure 5. Maps generated for Cedar Rapids, Iowa before (**left**) and after (**right**) mitigation using levees.

Model Generalization: Operational use and generalization capability of any flood map generation method is heavily dependent on the data availability and quality. While high-resolution DEM data (e.g., 1 m or 3 m) is not generally available around the world, many global elevation datasets [46] exist with moderate resolution (i.e., 30 m) and data quality. We tested HAND model at Cedar Rapids, Iowa (Figure 6) with low resolution elevation datasets (i.e., 10 m, 25 m, and 50 m) to understand the performance of the proposed system for limited data availability and quality. The resulting maps using 25 m or 50 m were relatively similar in flood map prediction quality compared to maps based on 10 m resolution datasets.

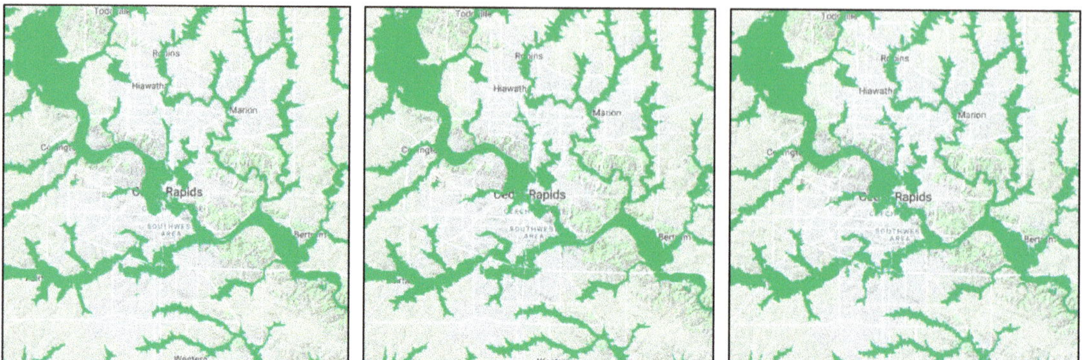

Figure 6. Maps generated for Cedar Rapids, Iowa with HAND model using 10 m, 25 m, and 50 m (left to right).

Low resolution elevation dataset caused small percentage (<5%) of the areas to be overestimated and underestimated compared to high resolution maps due to changes in the connectivity of the terrain for flow direction calculations. This issue could be further reduced while creating different resolution elevation datasets during resampling or surveying. The proposed HAND model includes terrain conditioning and clean up (i.e., pit removal and resolving flats) during flow direction generation. Low resolution elevation datasets can be further improved externally before connecting to the real-time system for more accurate results.

4. Conclusions

Real-time flood map generation is critical for operational activities in flood preparedness, mitigation, and response. Traditional flood mapping models requires extensive data and computation resources. In this study, we presented real-time flood map generation in client-side web systems using the HAND model with comparable results to FEMA-based reference maps. The height above the nearest drainage model computes the vertical distance from each cell to its nearest drainage. Testing of the HAND model in two key regions of Iowa demonstrated that the HAND model could be used to generate flood inundation predictions that are both accurate and fast. Predictions generated matched precomputed FEMA-approved flood insurance rate maps with acceptable accuracy within several seconds. Previous papers and applications of HAND focused on their minimal usage of external datasets, which allows for prediction of soil conditions and flood inundation extent in areas with little to no data aside from elevation datasets. The main application of our development is the ability of floodplain managers and end-users to add features to the terrain, such as levees and reservoirs, and examine how they will affect the extent of flooding. Since the users will be able to modify the terrain and see the result in real time, they will be able to get a better idea of how flood mitigation measures will affect flood inundation extents and be able to prepare for flooding in time with more accurate and relevant information. Future work can include an interactive user interface allowing selection of parameters, with potential automation and algorithms to select parameters to reflect real-world scenarios.

Author Contributions: Conceptualization, I.D.; methodology, A.H. and I.D.; software, A.H. and I.D.; validation, A.H.; formal analysis, A.H.; investigation, A.H. and I.D.; resources, A.H. and I.D.; data curation, A.H.; writing—original draft preparation, A.H. and I.D.; writing—review and editing, A.H. and I.D.; visualization, A.H. and I.D.; supervision, I.D.; project administration, I.D.; funding acquisition, I.D. All authors have read and agreed to the published version of the manuscript.

Funding: This research was supported by Secondary Student Training Program at the University of Iowa.

Institutional Review Board Statement: Not applicable.

Informed Consent Statement: Not applicable.

Data Availability Statement: Data used in this study is publicly available from Federal Emergency Management Agency, Iowa Flood Center and Iowa Department of Natural Resources.

Acknowledgments: The floodplain maps used in the analysis were acquired from the statewide mapping project at the Iowa Flood Center.

Conflicts of Interest: The authors declare no conflict of interest.

References

1. Coumou, D.; Rahmstorf, S. A decade of weather extremes. *Nat. Clim. Chang.* **2012**, *2*, 491–496. [CrossRef]
2. Noji, E.K. Natural Disasters. *Crit. Care Clin.* **1991**, *7*, 271–292. [CrossRef]
3. Bhola, P.K.; Leandro, J.; Disse, M. Framework for offline flood inundation forecasts for two-dimensional hydrodynamic models. *Geosciences* **2018**, *8*, 346. [CrossRef]
4. Tadesse, Y.B.; Fröhle, P. Modelling of Flood Inundation due to Levee Breaches: Sensitivity of Flood Inundation against Breach Process Parameters. *Water* **2020**, *12*, 3566. [CrossRef]
5. Arrighi, C.; Pregnolato, M.; Dawson, R.J.; Castelli, F. Preparedness against mobility disruption by floods. *Sci. Total Environ.* **2019**, *654*, 1010–1022. [CrossRef] [PubMed]
6. Bhatt, C.M.; Rao, G.S.; Diwakar, P.G.; Dadhwal, V.K. Development of flood inundation extent libraries over a range of potential flood levels: A practical framework for quick flood response. *Geomat. Nat. Hazards Risk* **2017**, *8*, 384–401. [CrossRef]
7. Singh, Y.K.; Dutta, U.; Prabhu, T.S.; Prabu, I.; Mhatre, J.; Khare, M.; Srivastava, S.; Dutta, S. Flood response system—A case study. *Hydrology* **2017**, *4*, 30. [CrossRef]
8. Yildirim, E.; Demir, I. An integrated web framework for HAZUS-MH flood loss estimation analysis. *Nat. Hazards* **2019**, *99*, 275–286. [CrossRef]
9. Yildirim, E.; Demir, I. An Integrated Flood Risk Assessment and Mitigation Framework: A Case Study for Middle Cedar River Basin, Iowa, US. *Int. J. Disaster Risk Reduct.* **2021**, *56*, 102113. [CrossRef]
10. Lamichhane, N.; Sharma, S. Development of flood warning system and flood inundation mapping using field survey and LiDAR data for the Grand River near the city of Painesville, Ohio. *Hydrology* **2017**, *4*, 24. [CrossRef]
11. Sermet, Y.; Demir, I. Towards an information centric flood ontology for information management and communication. *Earth Sci. Inform.* **2019**, *12*, 541–551. [CrossRef]
12. Sermet, Y.; Demir, I. Flood action VR: A virtual reality framework for disaster awareness and emergency response training. In Proceedings of the ACM SIGGRAPH 2019 Posters, Los Angeles, CA, USA, 28 July–1 August 2019; pp. 1–2.
13. Xiang, Z.; Demir, I. Distributed long-term hourly streamflow predictions using deep learning–A case study for State of Iowa. *Environ. Model. Softw.* **2020**, *131*, 104761. [CrossRef]
14. Dettinger, M. Climate change, atmospheric rivers, and floods in California—A multimodel analysis of storm frequency and magnitude changes. *JAWRA J. Am. Water Resour. Assoc.* **2011**, *47*, 514–523. [CrossRef]
15. Gaál, L.; Szolgay, J.; Kohnová, S.; Parajka, J.; Merz, R.; Viglione, A.; Blöschl, G. Flood timescales: Understanding the interplay of climate and catchment processes through comparative hydrology. *Water Resour. Res.* **2012**, *48*, W04511. [CrossRef]
16. Di Baldassarre, G.; Uhlenbrook, S. Is the current flood of data enough? A treatise on research needs for the improvement of flood modelling. *Hydrol. Process.* **2012**, *26*, 153–158. [CrossRef]
17. Sermet, Y.; Villanueva, P.; Sit, M.A.; Demir, I. Crowdsourced approaches for stage measurements at ungauged locations using smartphones. *Hydrol. Sci. J.* **2020**, *65*, 813–822. [CrossRef]
18. Seo, B.C.; Keem, M.; Hammond, R.; Demir, I.; Krajewski, W.F. A pilot infrastructure for searching rainfall metadata and generating rainfall product using the big data of NEXRAD. *Environ. Model. Softw.* **2019**, *117*, 69–75. [CrossRef]
19. Ebert-Uphoff, I.; Thompson, D.R.; Demir, I.; Gel, Y.R.; Karpatne, A.; Guereque, M.; Kumar, V.; Cabral-Cano, E.; Smyth, P. A vision for the development of benchmarks to bridge geoscience and data science. In Proceedings of the 17th International Workshop on Climate Informatics, Boulder, CO, USA, 20–22 September 2017.
20. Teng, J.; Jakeman, A.J.; Vaze, J.; Croke, B.F.; Dutta, D.; Kim, S. Flood inundation modelling: A review of methods, recent advances and uncertainty analysis. *Environ. Model. Softw.* **2017**, *90*, 201–216. [CrossRef]
21. Sit, M.; Demiray, B.Z.; Xiang, Z.; Ewing, G.J.; Sermet, Y.; Demir, I. A comprehensive review of deep learning applications in hydrology and water resources. *Water Sci. Technol.* **2020**, *82*, 2635–2670. [CrossRef]
22. Esfandiari, M.; Abdi, G.; Jabari, S.; McGrath, H.; Coleman, D. Flood Hazard Risk Mapping Using a Pseudo Supervised Random Forest. *Remote Sens.* **2020**, *12*, 3206. [CrossRef]
23. Priestnall, G.; Jaafar, J.; Duncan, A. Extracting urban features from LiDAR digital surface models. *Comput. Environ. Urban Syst.* **2000**, *24*, 65–78. [CrossRef]

24. Sinha, S.; Sotiropoulos, F.; Odgaard, A. Three-dimensional numerical model for flow through natural rivers. *J. Hydraul. Eng.* **1998**, *124*, 13–24. [CrossRef]
25. McGrath, H.; Bourgon, J.F.; Proulx-Bourque, J.S.; Nastev, M.; El Ezz, A.A. A comparison of simplified conceptual models for rapid web-based flood inundation mapping. *Nat. Hazards* **2018**, *93*, 905–920. [CrossRef]
26. Barthélémy, S.; Ricci, S.; Rochoux, M.C.; Le Pape, E.; Thual, O. Ensemble-based data assimilation for operational flood forecasting–On the merits of state estimation for 1D hydrodynamic forecasting through the example of the "Adour Maritime" river. *J. Hydrol.* **2017**, *552*, 210–224. [CrossRef]
27. Schumann, G.; Bates, P.D.; Apel, H.; Aronica, G.T. Global flood hazard mapping, modeling, and forecasting: Challenges and perspectives. In *Global Flood Hazard: Applications in Modeling, Mapping, and Forecasting*; John Wiley & Sons: Hoboken, NJ, USA, 2018; pp. 239–244.
28. Wu, W.; Rodi, W.; Wenka, T. 3D numerical modeling of flow and sediment transport in open channels. *J. Hydraul. Eng. ASCE* **2000**, *126*, 4–15. [CrossRef]
29. Afshari, S.; Tavakoly, A.A.; Rajib, M.A.; Zheng, X.; Follum, M.L.; Omranian, E.; Fekete, B.M. Comparison of new generation low-complexity flood inundation mapping tools with a hydrodynamic model. *J. Hydrol.* **2018**, *556*, 539–556. [CrossRef]
30. Gallant, J.C.; Dowling, T.I. A multiresolution index of valley bottom flatness for mapping depositional areas. *Water Resour. Res.* **2003**, *39*, 1347. [CrossRef]
31. Renno, C.D.; Nobre, A.D.; Cuartas, L.A.; Soares, J.V.; Hodnett, M.; Tomasella, J.; Saleska, S. HAND, a new terrain descriptor using SRTM-DEM: Mapping terra-firme rainforest environments in Amazonia. *Remote Sens. Environ.* **2008**, *112*, 3469–3481. [CrossRef]
32. Nobre, A.D.; Cuartas, L.A.; Hodnett, M.; Renno, C.D.; Rodrigues, G.; Silveira, A.; Waterloo, M.; Saleska, S. Height above the nearest drainage—A hydrologically relevant new terrain model. *J. Hydrol.* **2011**, *404*, 13–29. [CrossRef]
33. Nobre, A.D.; Cuartas, L.A.; Momo, M.R.; Severo, D.L.; Pinheiro, A.; Nobre, C.A. HAND contour: A new proxy predictor of inundation extent. *Hydrol. Process.* **2016**, *30*, 320–333. [CrossRef]
34. Li, Z.; Mount, J.; Demir, I. Model Parameter Evaluation and Improvement for Real-Time Flood Inundation Mapping Using HAND Model: Iowa Case Study. *EarthArxiv* **2020**. [CrossRef]
35. Xu, H.; Demir, I.; Koylu, C.; Muste, M. A web-based geovisual analytics platform for identifying potential contributors to culvert sedimentation. *Sci. Total Environ.* **2019**, *692*, 806–817. [CrossRef]
36. Sit, M.; Langel, R.J.; Thompson, D.; Cwiertny, D.M.; Demir, I. Web-based data analytics framework for well forecasting and groundwater quality. *Sci. Total Environ.* **2020**, *761*, 144121. [CrossRef] [PubMed]
37. Agliamzanov, R.; Sit, M.; Demir, I. Hydrology@ Home: A distributed volunteer computing framework for hydrological research and applications. *J. Hydroinform.* **2020**, *22*, 235–248. [CrossRef]
38. Sit, M.; Sermet, Y.; Demir, I. Optimized watershed delineation library for server-side and client-side web applications. *Open Geospat. Data Softw. Stand.* **2019**, *4*, 1–10. [CrossRef]
39. Haas, A.; Rossberg, A.; Schuff, D.L.; Titzer, B.L.; Holman, M.; Gohman, D.; Wagner, L.; Zakai, A.; Bastien, J.F. Bringing the web up to speed with WebAssembly. In Proceedings of the 38th ACM SIGPLAN Conference on Programming Language Design and Implementation, Barcelona, Spain, 18–23 July 2017; ACM: New York, NY, USA, 2017.
40. Sermet, Y.; Demir, I.; Muste, M. A serious gaming framework for decision support on hydrological hazards. *Sci. Total Environ.* **2020**, *728*, 138895. [CrossRef] [PubMed]
41. Xu, H.; Windsor, M.; Muste, M.; Demir, I. A web-based decision support system for collaborative mitigation of multiple water-related hazards using serious gaming. *J. Environ. Manag.* **2020**, *255*, 109887. [CrossRef] [PubMed]
42. Rebolho, C.; Andréassian, V.; Moine, N.L. Inundation mapping based on reach-scale effective geometry. *Hydrol. Earth Syst. Sci.* **2018**, *22*, 5967–5985. [CrossRef]
43. Barnes, R.; Lehman, C.; Mulla, D. Priority-flood: An optimal depression-filling and watershed-labeling algorithm for digital elevation models. *Comput. Geosci.* **2013**, *62*, 117–127. [CrossRef]
44. Barnes, R.; Lehman, C.; Mulla, D. An efficient assignment of drainage direction over flat surfaces in raster digital elevation models. *Comput. Geosci.* **2013**, *62*, 128–135. [CrossRef]
45. Gilles, D.; Young, N.; Schroeder, H.; Piotrowski, J.; Chang, Y.J. Inundation mapping initiatives of the Iowa Flood Center: Statewide coverage and detailed urban flooding analysis. *Water* **2012**, *4*, 85–106. [CrossRef]
46. Abrams, M.; Crippen, R.; Fujisada, H. ASTER Global Digital Elevation Model (GDEM) and ASTER Global Water Body Dataset (ASTWBD). *Remote Sens.* **2020**, *12*, 1156. [CrossRef]

Article

Flood Risk Communication Using ArcGIS StoryMaps

Khalid Oubennaceur *, Karem Chokmani, Anas El Alem and Yves Gauthier

Institut national de la recherche scientifique-Centre Eau Terre Environnement, 490 De la Couronne Street, Quebec City, QC G1K 9A9, Canada; Karem.chokmani@inrs.ca (K.C.); Anas.ELAlem@inrs.ca (A.E.A.); Yves.Gauthier@inrs.ca (Y.G.)
* Correspondence: khalid.oubennaceur@inrs.ca; Tel.: +1-418-999-0861

Abstract: In Canada, flooding is the most common and costly natural hazard. Flooding events significantly impact communities, damage infrastructures and threaten public security. Communication, as part of a flood risk management strategy, is an essential means of countering these threats. It is therefore important to develop new and innovative tools to communicate the flood risk with citizens. From this perspective, the use of story maps can be very effectively implemented for a broad audience, particularly to stakeholders. This paper details how an interactive web-based story map was set up to communicate current and future flood risks in the Petite-Nation River watershed, Quebec (Canada). This web technology application combines informative texts and interactive maps on current and future flood risks in the Petite-Nation River watershed. Flood risk and climate maps were generated using the GARI tool, implemented using a geographic information system (GIS) supported by ArcGIS Online (Esri). Three climate change scenarios developed by the Hydroclimatic Atlas of Southern Quebec were used to visualize potential future impacts. This study concluded that our story map is an efficient flood hazard communication tool. The assets of this interactive web mapping tool are numerous, namely user-friendly mapping, use and interaction, and customizable displays.

Keywords: story maps; disaster risk reduction; flood risk; slide; GARI tool; risk communication; climate change

Citation: Oubennaceur, K.; Chokmani, K.; El Alem, A.; Gauthier, Y. Flood Risk Communication Using ArcGIS StoryMaps. *Hydrology* **2021**, *8*, 152. https://doi.org/10.3390/hydrology8040152

Academic Editors: Marina Iosub and Andrei Enea

Received: 8 September 2021
Accepted: 7 October 2021
Published: 11 October 2021

Publisher's Note: MDPI stays neutral with regard to jurisdictional claims in published maps and institutional affiliations.

1. Introduction

Flooding is the most common and costly natural hazard affecting Canadians today [1]. Floods occur throughout the country, but most of the worst flooding events have happened in southern regions, affecting hundreds of thousands of people and costing billions of dollars in damage [2]. In addition, scientists expect an increase of flooding events and damage costs in riparian, coastal, and urban areas of the country [3]. Flooding poses a serious risk to households and communities across Canada. In the spring of 2019, for instance, flooding in eastern Canada cost more than CAD 200 million in insured losses and damage to nearly 20,000 properties [4]. Despite such impacts, Canadians are largely unaware of the flood risk they face. According to a survey published in 2016 based on 2300 households by the Interdisciplinary Center on Climate Change by Partners for Action at the University of Waterloo [4], only 6% of Canadians living in a designated flood risk area are aware of the risk, and only 21% believe that the risk of flooding will increase over the next 25 years. Furthermore, Canadians need and want more information on how to be actively engaged in flood management and how to improve their resilience. In this context, efforts are being made to reduce the impacts of flooding events by increasing public awareness through communication of flood hazards and popularization of flood risk reduction measures.

Communication plays an important role in flood risk management since it has the power to raise risk awareness on potential flood hazards, educate and motivate the population at risk to take preventive actions, and to be prepared in the case of an emergency. Information tools in flood risk communication aim to inform the people about risks in their residential zone. Various research projects have been conducted to better understand the mechanisms of effective flood hazard communication [5–8]. Mees et al. [9], for example, have demonstrated the need for accessible communication tools to raise awareness of flood and climate change impacts. The use of flood risk maps to show the potential severity of floods and locations at risk has been shown to improve citizen response to floods in the United Kingdom [10]. Flood maps can be used to different ends, such as hazard and risk identification and land use planning. They can be used for many aims and by various users, such as identifying which properties should adopt flood-proofing measures and educating the public about local flood risks. Different forms of flood maps exist and it is important to distinguish flood hazard maps from flood risk maps. Flood hazard maps [11] show flood depth and water extent for given events, whereas flood risk maps [12] indicate the potential consequences of floods on properties and social, economic, environmental, and cultural assets, along with the global risk faced by a community to certain flood scenarios. De Moel et al. [13] highlighted that flood hazard maps are the most common type of flood maps, followed by flood risk maps.

The use of online interactive communication tools such as story maps to dispense information is becoming increasingly popular. Such web applications facilitate communication by illustrating complex scientific data in an accessible format to specific target groups. Besides being informative for target users, story maps should also be easy to understand and use. Web-based story maps can be static or dynamic. While static webmaps are based on fixed images, dynamic webmaps can be adjusted according to the user's preferences, which can select different scenarios or zoom over a desired location, for example. Both map types offer the possibility of choosing the information displayed on the map (e.g., flood hazard, flood risk maps, etc.). Recent advances in internet accessibility and speed have led to an increased interest in favour of using webmaps as a mode of visual communication.

Story mapping represents a novel mixed media approach that combines different functionalities such maps, videos, images, graphs, and text into a simple online interface. New and innovative GIS tools such as ArcGIS Online and ArcGIS StoryMaps (Esri) are much more than thematic maps; they are a complete and promising means of communication. A major advantage is that story maps can be used by non-GIS users to showcase and popularize the results of scientific research and that is partly why they represent a good option in flood hazard and risk communication. Individuals can be informed of their flood risk level while actually visualizing their houses [14]. They allow the combination of mapped flood hazard and risk data with other information to enhance both the map's content and the quality of flood hazard and risk communication. The tools available within ArcGIS StoryMaps help create interactions with flood risk information that goes beyond what can be achieved by flood risk maps alone. Story maps can illustrate the spatiality of flood risks supplemented with critical information, informing users on numerous flooding-related topics. In Canada, the city of Calgary has created a story map for flood risk and resilience: https://maps.calgary.ca/RiverFlooding/ (accessed on 1 September 2021), which contains images, text, flood maps, and testimonies from flood victims. This application allows users to look up their individual address and see if they are within the floodway zone. Furthermore, story maps can be used in climate change studies to communicate potential flooding impacts [15–17].

Flood risk maps are one of the visual components of a story map and should be developed with an easy-to-understand content, a user-friendly interface, and links to further information [18]. Story map design plays a crucial role in the interpretation, so elements such as legends, colours, embedded pop-ups, and technical terminology should be used to enhance user experience. For example, flood risk map legends should draw the user's attention by being clear, clearly displayed, simple to understand (low−medium−high risk levels, shades of blue for flood extents, and avoiding return period terminology, rather frequent−medium−extreme events) and contain a conservative amount of information [19]. The category classes also have to be comprehensible and readable at first sight. Furthermore, the contrast of the map's background containing informative elements should be optimized to prevent information overload [19].

Given the benefits of story maps to crisis managers and the public, in risk communication, it was decided to develop one as a case study for the Petite-Nation River. The main elements of this interactive story map included the flood risk of previously flooded areas, the buildings and streets affected during different flooding events, climate projection maps, and information about past flooding events in the Petite-Nation River watershed. This initiative was aimed at assessing the potential of story mapping to disseminate scientific data and relevant information to the general public, with regard to flooding management. For this case study, ArcGIS software by Esri was used alongside ArcGIS Online and the ArcGIS StoryMaps story builder.

2. Materials, Methods, and Application Case

2.1. Study Area

The study area is located in the Petite-Nation River watershed in Quebec, Canada (Figure 1). The watershed covers an area of 2248 km^2. This sector has been regularly experiencing flood events due to its location on the Petite-Nation River floodplain. The most recent event occurred in 2019 and affected the municipalities of St-Andre-Avellin and Duhamel. Many roads were flooded. The municipality of St-Andre-Avellin declared afterwards a state of emergency resulting in the evacuation of approximately 100 people.

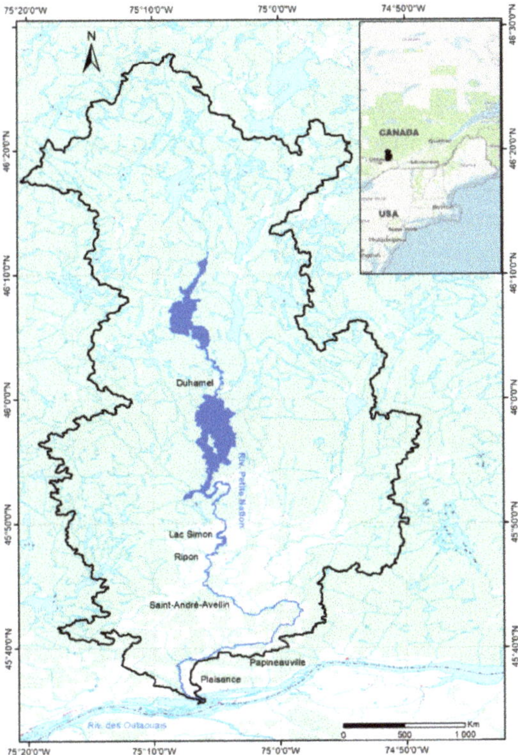

Figure 1. The Petite-Nation watershed study area.

This study is part of a larger project named FRAPNR (flood risk assessment on the Petite-Nation River watershed), which is carried out by the National Institute of Scientific Research (INRS), in partnership with the local watershed organization (Organisme de bassins versants des rivières Rouge, Petite-Nation et Saumon; OBVRPNS) and six municipalities located within the watershed (Duhamel, Ripon, St-Andre-Avellin, Lac-Simon, Papineauvillle, and Plaisance). This project aims at improving community response to flooding by increasing the understanding of the impacts of climate change, promoting upstream−downstream solidarity among municipalities, and helping vulnerable communities to be better equipped to deal with climate change-related flooding [20]. For this case-study, we designed a story map in collaboration with the partners to address flooding and climate change. To effectively do this, we also produced and included a series of flood maps for the watershed of the Petite-Nation River.

Meetings with different stakeholders of the Petite-Nation River watershed were held to better identify user needs and to address the best ways in which they could utilize the developed story maps as part of their regular responsibilities. The consulted stakeholders included municipal representatives such as the town manager, community planners, public works manager, emergency planning team, officers of the environment department, and citizens. The aim of these meetings was to take stock of their opinions on current flooding-related issues, assess suitable communication methods, gather their views in regard to the use of story maps, and to assess their potential as flood hazard and risk communication resources.

2.2. The GARI Tool

In order to visualize the extent of flooding events under different climate change scenarios, we developed a series of flood risk maps using GARI [21,22]. The GARI tool contains three modules described as follows:

Flood mapping: shows the flooding extent and depth delineation on a basemap. The flood hazard maps used in this case study were produced using the HAND method (height above nearest drainage). The HAND technique is a low-complexity, field-based approach for inundation mapping using elevation data, discharge–height relationships, and stream discharge flows. Producing a HAND map requires a digital elevation model (DEM) and a spatial representation of a region's river network. The HAND technique computes the elevation difference between each land surface cell and the stream bed cell to which it drains. The resulting HAND map contains the vertical distance between a location and its nearest stream [23].

Vulnerability of the population: indicates the vulnerability and potential adverse consequences resulting from flooding events including social and environmental aspects. Three different shapefiles can be produced using this module: the first one presents the level of risk to the population per building (flood exposure), the second contains the flooded and nonflooded critical infrastructure, and the third illustrates the flooded road sections. The vulnerability of the population is estimated and mapped for every building. This module also includes water height at flooded critical infrastructures and flooded road portions. The first layer, which is the exposure of the population to a flood hazard can be determined by using the characteristics of the buildings in which the citizens are located at the time of the flood. The second layer, which contains information on the building, includes the following information: registration number, civic address, state of the building, maximum water level at building level, socio-economic vulnerability of the residents, exposure to flood hazard, and the estimated final risk level for the residents (low, serious, severe, and very severe). The attributes of the third layer are the road number, name, condition with regard to water level (nonflooded, passable (< 25cm), emergency vehicles only (25 to 50 cm), impassable), and the maximum water level on each flooded road section.

Damage to buildings and infrastructure: this module generates a map of the estimated damage to each building during a given flood event. This damage is estimated and mapped for each building and expressed in absolute (CAD) or relative (%) terms of the building value. For this purpose, four damage curves (A damage curve indicates the level of damage to a residence as a function of the water height at the main floor level. There is a specific curve for each type of residence) covering an area encompassing one thousand buildings (appx.) was used. A damage curve indicates the level of damage to a residence as a function of the water height at the main floor level. For each residential building, the software estimates the damage based on the water height in the building, by using the curve adapted to its structural characteristics. The total damage to the study area is the sum of each residential building's damage value.

The outputs of each module of the GARI tool are summarized in Figure 2. The GARI modelling outputs were used to develop the story map presented in this case study. These outputs were imported into GIS layers using ArcMap, then uploaded as an image service into ArcGIS Online, and finally implemented as story maps.

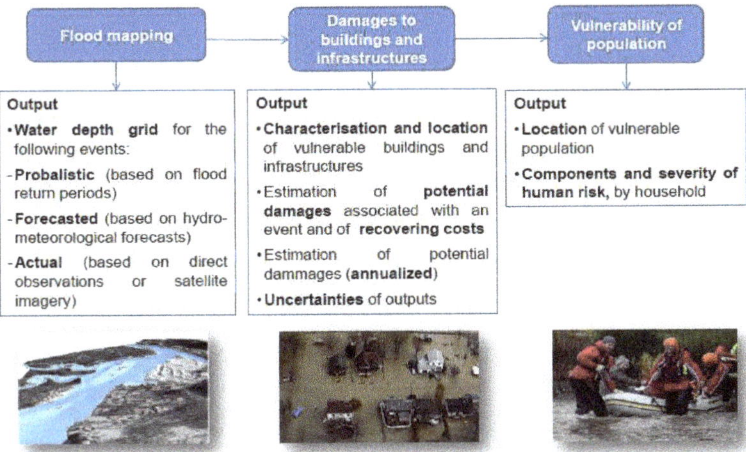

Figure 2. Output information for all three modules of the GARI tool.

2.3. Available Data

Different types of datasets were selected to set up the flood and climate risk story map of the Petite-Nation River watershed (Table 1). The data falls under two main components of flood risk: hazard and vulnerability (Figure 3). Flood risk is defined as a function of flood hazard and the vulnerability of communities to a flood with a particular return period.

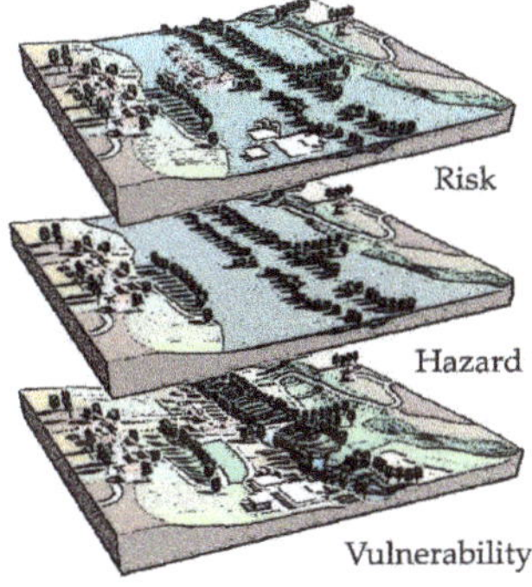

Figure 3. Flood risk as the interaction of hazard and vulnerability, source:(http://www.prim.net, accessed on 1 September 2021).

The field values (attributes) and the corresponding variable format of the vector data used to map the flood risk and climate change are shown in Table 2. The raster data corresponds to the depth of water in the pixel.

Table 1. Data used for the construction of the story map.

Module	Input Data	Scale	Format	Source	Output
Vulnerability of population	Most recent property assessment roll (2017)	Building	Tabular (.xls format)	Municipality (upon request)	Vector vulnerability (socio-economic vulnerability + people exposure to flooding + vulnerability due to lack of essential resources)
	2016 Census; dissemination area boundaries	Dissemination area	Tabular (.ivt format) and polygon(.shp format) for the boundaries	Statistics Canada Census (online)	
	Shape of the building	Building	Polygon (.shp format)	Municipality (upon request)	
	Cadastral map	Cadastral lot	Polygon (.shp format)	Municipality (upon request)	
	Road network (206)	Scale 1:20 000	Vector (.shp format)	National Topographic DataBase, Natural Resources Canada (NRCAN)	
Flood mapping	Water depth	Pixel (1 m)	(.tif or .grid format for depth) and (.shp for extent)	INRS (HAND Method)	Water depth raster (grid)
	Water marks for validation	Pixel (1 m)	Point (.shp)	INRS	
	DEM (Digital Elevation Model)	1 m	(.tif or .grid format)	Ministère des Forêts, de la Faune et des Parcs	
	Hydrometric data (discharges)	Station	Tabular (.xls)	CEHQ (hydrometric station at Ripon)	
	Hydrographic river network		.gdb geodatabase		
Damage to buildings	Dissemination area boundaries 2016	Dissemination Area	(.shp format)	Statistics Canada (online)	Building damage vector (shapefile)
	Flood damage curves	Tabular	Tabular (.xls format)	INRS	
	Most recent property assessment roll	Building	Tabular (.xls format)	Municipality (upon request)	
	Shape of the building	Building	Polygon (.shp format)	Municipality (upon request)	
	Most recent cadastral map (optional)	Cadastral lot	Polygon (.shp format)	Municipality (upon request)	
	Water depth	Pixel (1 m)	(.tif or .grid format for the depth) and (.shp for the extent)	INRS (HAND method)	
	First floor height	Meter (1 m)	tabular (.xls format)	Municipality (on request)	

Table 2. Field values and variable format used to develop the story map.

Field	Variable format
Vulnerability of population: Flood exposure	
Address of the building	String
State of the building: normal, isolated, building affected, flooded basement, basement with first floor flooded, flooded first floor (without basement).	Long
Vulnerability of population: Road network	
Name of the flooded road segment	String
Status of the road segment: nonflooded, flooded (impassable), flooded (passable), flooded (emergency vehicles only)	String
Length of the segment	Double
Damage to buildings	
Registration number	Long
Damage (%)	Float
Damage ($)	Float

2.4. *Generation of Flood Risk and Climate Change Maps*

2.4.1. Flood Risk Maps

Nine flood risk maps were developed based on the scenarios shown in Table 3. These maps were generated by combining the flood hazard maps with population vulnerability maps. The flood hazard was obtained through the HAND method and composed of the raster files (water depth grids) associated with different discharges. The estimated discharge values for the Petite-Nation River watershed correspond to the discharge flow at the hydrometric station located in Ripon. Threshold discharge values are defined by Quebec's Ministère de la Sécurité Publique (department of public security). The flood hazard maps are mainly composed of the raster files (water depth grids). These files were gradually transformed into a format suitable for ArcGIS Online (feature or image service) and included in the story map. The population vulnerability was based on the vector of the flood exposure and the flooded road network, generated by the module vulnerability of population of GARI tool. The combination of the hazard and vulnerability allowed a complete and relevant estimate of the flood risk to be obtained. The flood risk maps were estimated and mapped for each residential building present in the study.

2.4.2. Flood Risk Maps

In addition to the flood risk maps, three maps based on climate change projections were also created. Climate change projections were derived from the Hydro-Climatic Atlas of Southern Quebec issued by the Coupled Model Intercomparison Project phase 5 (CMIP 5) [24] following the pessimistic Representative Concentration Pathways (RCPs) scenarios RCP 8.5. This scenario was selected because it is the one most commonly employed for impact assessments. The details of these climate projections are summarized in Table 4. The projections are based on the estimated development for 2050 and 2080. Present and future flood hazard and vulnerability will help stakeholders better understand the potential impact of climate change on flood risk in the Petite-Nation River watershed. These climate change maps were created for the two most populated municipalities of the Petite-Nation River watershed (Ripon and St-Andre-Avellin) and included projected flood hazard and monetary damage layers for the baseline, 2050 and 2080 horizons. The monetary damage was estimated using the GARI software (damage to buildings and infrastructure module).

Table 3. The nine scenarios for the Petite-Nation River watershed with their corresponding discharge expressed in m^3/s and return period in years.

Scenario	Estimated Discharge (m^3/s)	Return Period (Years)	Flooding Threshold Level
Scenario 1	96	2	Minor
Scenario 2	115	5	Minor
Scenario 3	128	10	Medium
Scenario 4	139	20	Medium
Scenario 5	149	35	Medium
Scenario 6	154	50	Major
Scenario 7	159	75	Major
Scenario 8	163	100	Major
Scenario 9	213	Up to 200	Major (Spring 2019)

Table 4. Climate change projections for the Petite-Nation River watershed using a return period of 20 years for the baseline and the 2050 and 2080 horizons (RCP 8.5).

Climate Change Scenarios	Horizon (Years)	Estimated Discharge (m^3/s)
Scenario 10	Baseline	139
Scenario 11	2050	132
Scenario 12	2080	127

2.5. Story Map Template

The geospatial web technology selected in this study is the ArcGIS StoryMaps application. This application can be created using several designs that need no coding, or creators can develop their own StoryMap designs [25]. A story map slide for each location in the story: the location can be inserted by performing a text search for the name, address, etc. When viewing the story map slides, the side bar on the left helps to follow the narrative, and the majority of the screen is taken up by a map, which may be panned and zoomed. There are several story mapping templates included in the ArcGIS StoryMaps application, namely Story Map Tour, Journal, Cascade, Shortlist, Swipe, Basic, and Series [26]. The template used in this study is from the StoryMap Series (Figure 4).

Figure 4. Example of a story map developed with ArcGIS StoryMap Series—Title: Calgary's River Flood Story (https://maps.calgary.ca/RiverFlooding/, accessed on 1 September 2021).

The StoryMap Series allows the assembling of a series of maps via tabs, numbered bullets, or an expandable side accordion using an interactive builder. The pages of the story (called map sheets) can be designed by using a template in which each page shows a particular map. In addition to maps, images, and videos, web content can also be added to create a narrative and engage your audience. For the Petite-Nation River case study, we selected the tabs template (https://arcg.is/OvP98, accessed on 1 September 2021).

2.6. Story Map Construction

The first step of the story map construction process was to proceed to geospatial data structuration and their spatial and descriptive analyses. The ArcGIS platform was used to collect, analyze, and visualize data. The vector data are mainly shapefiles of flood risk exposure (polygon), damage to buildings (polygon), damage to the road network (polyline), and the raster data is the water depth. A geodatabase was created in ArcMAP 10.7.1 (Esri) to stock the information layers that would also appear on the online story map. Each information layer hosts one type of spatial information along with the necessary fields for the descriptive information (Table 5). Then, the collected files were uploaded (vector and raster) to ArcGIS Online platform in order to construct the online map (webmap). These files were gradually transformed into a format suitable for online use (feature or image services) and uploaded directly to the online platform. Vector data were zipped and directly imported into ArcGIS Online, and the raster data were converted into image service and then imported into ArcGIS Online. Images were converted to the Flickr format in order to be compatible with the accepted template formats. Layer parameters (e.g., symbol, transparency, class, tag display, and pop-up menus) were then created. Specific legends and symbols were also created for every data type. The legend used for flood risk mapping includes three components: water depth, risk exposure of buildings, and risk exposure of the road network. For the water depth, a blue-color palette was used. For the building and roads risk data, different colors were used to differentiate building status (from green for normal buildings, to red for buildings with a basement and first floor flooded). The categorization and symbolization of roads (from green for nonflooded segments to red for flooded roads authorizing only emergency vehicles) was added.

Table 5. Story map slide design.

Slide #	Content and Sequential Order of Information
1	Disclaimer Medium: image explaining how to use this story map
2	Project partners Medium: image showing project partners
3	Project context Right: image of a flooded sector of St-Andre-Avellin, spring 2019 Left: text information regarding flooding in the Petite-Nation River watershed, the aim of this story map and an image showing the study area
4	GARI tool Right: diagram of the three main components of GARI tool used to simulate flood risk maps under specific scenarios for slides 6-17 Left: text description of the GARI tool. These terms are important to understand the flood risk and climate change maps (the following slides).
5	Basic definition of flood risk Right: image of the three flood risk components: hazard, vulnerability and assets. Left: description of each flood risk component
6-14	Flood risk maps corresponding to 9 different scenarios displayed from the lowest discharge flow (96 m^3/s) to the highest (213 m^3/s) Right: ArcGIS map of flood risk Left: description of each scenario followed by the flood risk legend (water depth, and roads/building state)

Table 5. *Cont.*

Slide #	Content and Sequential Order of Information
14-17	Flood risk maps based on climate change projections (baseline, 2050, 2080 using the RCP 8.5 and a 20-year return period Right: ArcGIS flood risk map Left: description of each scenario followed by the flood hazard legend (water depth, risk for roads, and damage to building)

The transparency of raster data was set at 50%. The narrative text combined with the maps was kept as brief as possible, especially for map slides. For each webmap, we included a short text explaining the concepts of discharge flow value, return period, and thresholds set up by the public security department. Only the most important information and attributes (Table 2) are displayed to the user to avoid clutter and information overload (from ESRI's basic gallery maps). Finally, two basemaps available from ArcGIS Online were assigned as the background: Google Earth Webmaps were used for the flood risk scenarios and topographic webmaps were used for the climate change scenarios. Thematic maps were created in ArcGIS Online based on the collected data.

The story map presented in this paper is composed of 17 slides (or tabs) through which the users can browse. Each slide is divided in two: the left part contains the narrative text, photos, and tables, and the right part which is the main frame, displays the webmaps, scenes, or multimedia content. The first four slides provide a general description of the project and the study area, the following nine slides contain the flood risk maps, and the next three slides present the climate change projections (Table 5). Layout and content variations were minimal between slides. Table 5 illustrates the content and design of the seventeen story map slides.

3. Results

3.1. Online Flood Risk Story Maps

The information contained in this story map was organized to present the flood risk over the Petite-Nation River watershed based on increasingly severe discharge scenarios (slides 6−14). When the discharge flow increases, the risk, hazard, number of flooded buildings, and roads also increase. A total of nine scenarios related to nine water discharge flows ranging from 96 m^3/s (2-year period) to 213 m^3/s (maximum flow recorded in the spring of 2019 (Table 3)). The selected discharge flow values cover the full range of flood conditions in the watershed of the study (https://arcg.is/OvP98, accessed on 1 September 2021).

Figure 5 provides representative screenshots of slide designs 6, 9, and 14. These scenarios correspond to minor, medium and major discharge flows: 96 m^3/s (2 years), 139 m^3/s (20 years), and 213 m^3/s (up to 200 years), respectively. For each of these slides, in the left side of the interface, the users can find information about discharge values, return period used in this scenario with its threshold, and just below the legend used for the flood risk map. This legend contains: intensity of water depth, building exposure status, and road status in terms of flooding. On the right side of the slide, the user can explore and visualize the web flood risk map corresponding to a given scenario. Users, by a simple click of buttons, can zoom in and out, move the map, identify the status (with its displayed color), the address of the building, and the status of road segment flooded and its length. The colors on this map show areas at risk of river flooding. The high resolution of the maps would allow decision makers and emergency responders to take action at specific locations by knowing the intervention context.

Figure 5. Screenshot of slides 6 (**A**: Scenario 1), 9 (**B**: Scenario 4) and 14 (**C**: Scenario 9).

3.2. Climate Projection Online Story Maps

By changing the flood frequency analysis based on climate change projections with the associated inundated depth maps and the damage curves for the residential buildings and roads, we estimated the change in flood damage, expressed in percentage according to the value of the property for the horizons 2050 and 2080. Figure 6 shows representative screenshots of slide designs for slides 14, 15, and 16, which addressed the projected flood hazard and damage based on 20-year return periods for the municipalities of St-Andre-Avellin and Ripon under current and future horizons, assuming that there are no improvements or modifications to the existing flood protection network.

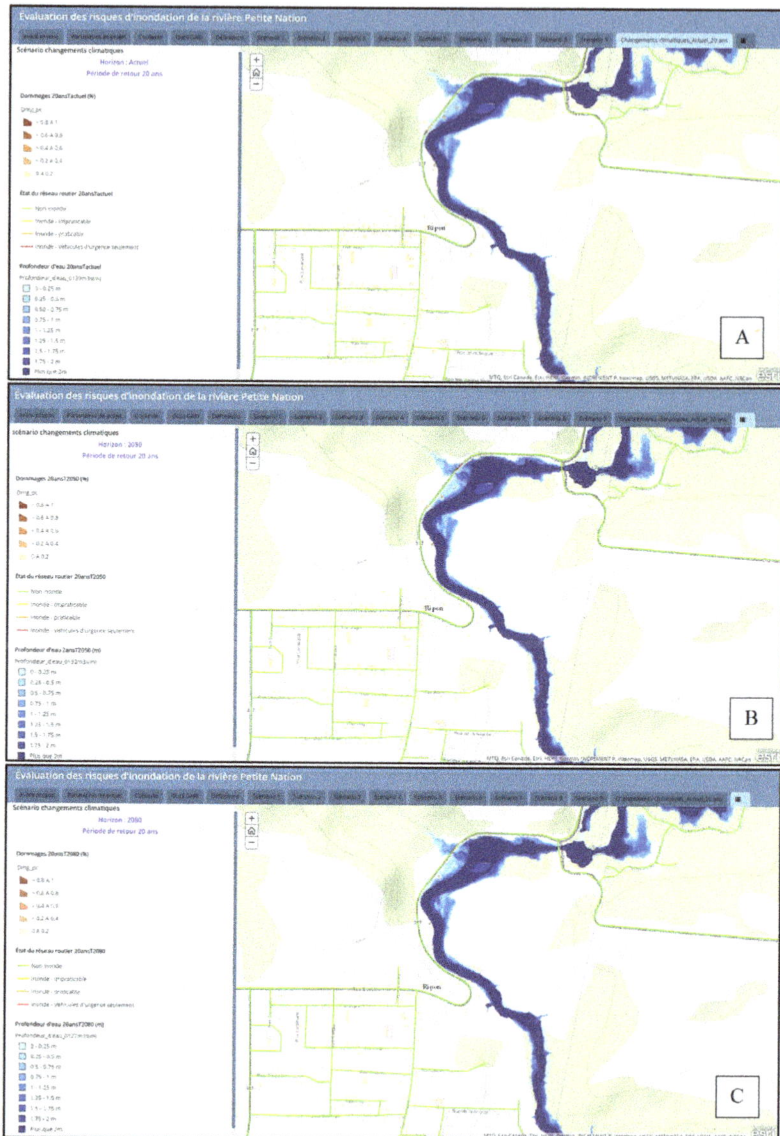

Figure 6. Screenshot of slides 14 (**A**: current with a 20-year return period), 15 (**B**: 2050 with a 20-year return period) and 16 (**C**: 2080 with a 20-year period) (https://arcg.is/OvP98, accessed on 1 September 2021).

The resulting webmaps illustrate flood damage at the scale of the building, combined with information on water depth. When navigating the slide of a given climate scenario, the user can explore and visualize the static risk map corresponding to the climate scenario by zooming into a desired location and clicking in a grid cell to view the projected change in flood hazard and damage. Information about climate change scenarios, the return period, and the RCP used in the slide, as well as the legend (water depth, building damage expressed in percentage, and the roads status) are displayed on the same screen to the left of map.

4. Discussion

In this work, we have developed a story map application for the Petite Nation Watershed based on flood risk and climate maps combined with informative texts. The study has been instructive in quantifying the potential of this web technology, its limitations and the critical influence.

The decision to use the StoryMaps application for this work was based on the numerous advantages compared to traditional methods:

1. The developed story map represents a flexible and interactive application to better understand the flooding in the Petite Nation River watershed.
2. This story map has made the flood risk approach a lot more transparent and dynamic, allowing the user to map flood risk under a range of scenarios. In addition, it demonstrates how narratives, specifically experiential stories from those in vulnerable areas can be combined with flood risk and climate projection maps to create an interactive, visual tool that gives more context to the risk.
3. It provides fast and inexpensive estimates of flood risk for specific return periods that complement results from hydrodynamic simulations.
4. The majority of the members of the municipalities of the Petite Nation watershed do not have any experience in manipulating geospatial datasets. Thus, the template-based approach of this story map was used in order to disseminate and make scientific findings easy-to-access and understand to broader nontechnical audiences. No GIS background or previous experience with coding or web development is needed.
5. The developed story map allows the combination of multiple different data (maps, images, narrative text, etc.) for better communication of information. This characteristic allows professional and nonprofessional users from various domains to easily understand the story maps.
6. The story map contains interactive, digital, and easy to use web maps that replace classical maps or analog maps printed on paper. The user can navigate easily through the content, by pop-ups, interact with and zoom around the maps, swipe up and down and through slides and scroll through the tabs in this story map to explore the impacts and consequences of flood and climate change on the Petite Nation watershed.
7. Story maps could be utilized as an education resource in schools, teaching children about flood risk in their community.
8. The flood risk web maps presented in this story map could also be used for: screening hotspots and elements at risk (e.g., buildings, critical infrastructures, etc.), confirming flood-prone locations with point information (e.g, flood depth), providing opportunities for individuals to further investigate their local area's flood risk, and understanding their own flood risk situation.
9. Beyond these technical considerations, the story map can also be used for educational purposes in schools in order to present concrete cases of GIS technique use to future generations of students.

However, the present state of this story map is affected by some limitations. This application at present is a simple mapping with reduced interactivity and little functionality, but with good readability and adequate explanations for the citizen. In general, the information communicated to the citizen by flood related authorities has to be adapted to the needs and requirements of the users to be as effective as possible. The currently existing flood risk maps still lack a good balance between simplicity and complexity. The best way for flood risk communication with the public is the use of real-time information such as gauge levels to inform about flood risks. Gauge levels are easier to understand than the return periods and allow the population at risk to compare these flood situations in the past event. Real flood events have a much stronger influence on building an awareness of flood risk perception than the flood events only derived from hydraulic modelling. In addition, the information communicated via these maps includes water depth with different returns periods, and lack information about floodplain extents. In the majority of cases of webmapping tools around the world, the extent of floods or a floodplain map is used as the basis and the most widely used information. Moreover, risk communication has to be easily comprehensible to the people at risk. This means avoiding technical terminology and statistical terms such as return periods.

5. Conclusions

In this study, a story map is proposed as a new method for communicating current flood risks and flood risk projections based on climate change scenarios. This proposition stems from the need to better popularize and communicate flood-related information to citizens. The method was applied to the Petite-Nation River watershed, which is regularly flooded. Geospatial data (vector and raster) were used and thematic maps were created depicting the flood risk under current conditions and climate change projections. These maps were the foundation of the story mapping process. The final story map was composed of slides addressing the study area, background information on flooding, interactive risk maps, and some context on the project and the development team.

New enhancements could further improve the StoryMap application developed in this study. For example, the addition of new dynamic functions such as formulating queries, search buttons or data selection would improve the user experience. Adding information about the extension of floodplains, and up-to-date water levels and forecasts (gauges levels or web cameras), and pictures of past events. Displaying the uncertainties related to the values displayed and the projections used would increase transparency, allowing the user to better interpret the data. Statistics such as the number of flooded areas and the affected streets and buildings could also be relevant information to include. Finally, the story map could be enriched by imbedding hyperlinks to interesting articles and websites that would support and provide opportunities for people to investigate and learn of their own flood risk situation.

If story maps were to be optimized for decision makers, such users could benefit from the addition of more indicators, such as socio-economic vulnerability. The quality of the information displayed in the story map could be enhanced by allowing dynamic flood forecasting using real hydrometric time data from the hydrometric station. A warning system could even be added leading to faster response delays. The webmap showing flood forecasts are similar in form to the present web maps, but their emphasis is on the future and they also need to explain the analysis on which the predictions are made. The combination of flood hazard maps with real time information on gauge levels is made possible by directly linking gauge stations with the flood forecast center.

Since the objective of this case study was to test the potential of story maps as a means of communication to citizens and decision makers residing in flood-prone areas, this story map focused on illustrating flood risk and future projections, while providing the information necessary to understand and interpret the maps. The use of specialized software developed by Esri, namely ArcMap, ArcGIS Online and ArcGIS story maps, has proven to be an efficient and effective way to develop the flood-related story map.

Author Contributions: Conceptualization, K.O.; methodology, K.O. and K.C.; writing—original draft preparation, K.O.; writing—review and editing, K.O., A.E.A.; visualization, Y.G., K.C.; validation, K.C., Y.G.; supervision, K.C. All authors have read and agreed to the published version of the manuscript.

Funding: This research was funded by the Municipalities for Climate Innovation Program led by the Federation of Canadian Municipalities, grant number MIC-15557.

Data Availability Statement: The data presented in this study are available on request from the first author Khalid Oubennaceur.

Acknowledgments: This study was carried out by the National Institute of Scientific Research (INRS), in partnership with the local watershed organization (Organisme de bassins versants des rivières Rouge, Petite-Nation et Saumon; OBVRPNS) and six municipalities located within the watershed (Duhamel, Ripon, St-Andre-Avellin, Lac-Simon, Papineauvillle, Plaisance), under the project entitled "Tool: Assess flood risk and develop sustainable management plans".

Conflicts of Interest: The authors declare no conflict of interest.

References

1. Buttle, J.M.; Allen, D.M.; Caissie, D.; Davison, B.; Hayashi, M.; Peters, D.L.; Pomeroy, J.W.; Simonovic, S.; St-Hilaire, A.; Whitfield, P.H. Flood processes in Canada: Regional and special aspects. *Can. Water Resour. J./Rev. Can. Ressour. Hydr.* **2016**, *41*, 7–30. [CrossRef]
2. MMM Group. *National Floodplain Mapping Assessment: Final Report*; Public Safety Canada: Ottawa, ON, Canada, 2014; Volume 68.
3. Winsemius, H.C.; Aerts, J.C.; Van Beek, L.P.; Bierkens, M.F.; Bouwman, A.; Jongman, B.; Kwadijk, J.C.; Ligtvoet, W.; Lucas, P.L.; Van Vuuren, D.P. Global drivers of future river flood risk. *Nat. Clim. Chang.* **2016**, *6*, 381–385. [CrossRef]
4. Thistlethwaite, J.; Henstra, D.; Peddle, S.; Scott, D.J. *Canadian Voices on Changing Flood Risk: Findings from a National Survey*; Action, P.F., Ed.; University of Waterloo: Waterloo, ON, Canada, 2017.
5. Bradford, R.; O'Sullivan, J.J.; Van der Craats, I.; Krywkow, J.; Rotko, P.; Aaltonen, J.; Bonaiuto, M.; De Dominicis, S.; Waylen, K.; Schelfaut, K.; et al. Risk perception–issues for flood management in Europe. *Nat. Hazards Earth Syst. Sci.* **2012**, *12*, 2299–2309. [CrossRef]
6. Faulkner, H.; Ball, D. Environmental Hazards and Risk Communication. *Environ. Hazards* **2007**, *7*, 71–78. [CrossRef]
7. O'sullivan, J.; Bradford, R.; Bonaiuto, M.; De Dominicis, S.; Rotko, P.; Aaltonen, J.; Waylen, K.; Langan, S. Enhancing flood resilience through improved risk communications. *Nat. Hazards Earth Syst. Sci.* **2012**, *12*, 2271–2282. [CrossRef]
8. Thieken, A.H.; Mariani, S.; Longfield, S.; Vanneuville, W. Preface: Flood resilient communities–managing the consequences of flooding. *Nat. Hazards Earth Syst. Sci.* **2014**, *14*, 33–39. [CrossRef]
9. Mees, H.; Tijhuis, N.; Dieperink, C. The effectiveness of communicative tools in addressing barriers to municipal climate change adaptation: Lessons from the Netherlands. *Clim. Policy* **2018**, *18*, 1313–1326. [CrossRef]
10. Ping, N.S.; Wehn, U.; Zevenbergen, C.; van der Zaag, P. Towards two-way flood risk communication: Current practice in a community in the UK. *J. Water Clim. Chang.* **2016**, *7*, 651–664. [CrossRef]
11. Nones, M. Flood hazard maps in the European context. *Water Int.* **2017**, *42*, 324–332. [CrossRef]
12. Oubennaceur, K.; Chokmani, K.; Nastev, M.; Lhissou, R.; El Alem, A. Flood risk mapping for direct damage to residential buildings in Quebec, Canada. *Int. J. Disaster Risk Reduct.* **2019**, *33*, 44–54. [CrossRef]
13. De Moel, H.; Van Alphen, J.; Aerts, J.J.N.H. Flood maps in Europe-methods, availability and use. *Nat. Hazards Earth Syst. Sci.* **2009**, *9*, 289–301. [CrossRef]
14. Frazier, T.G.; Wood, N.; Yarnal, B. Stakeholder perspectives on land-use strategies for adapting to climate-change-enhanced coastal hazards: Sarasota, Florida. *Appl. Geogr.* **2010**, *30*, 506–517. [CrossRef]
15. O'Neill, S.J.; Boykoff, M.; Niemeyer, S.; Day, S.A. On the use of imagery for climate change engagement. *Glob. Environ. Chang.* **2013**, *23*, 413–421. [CrossRef]
16. Spence, A.; Pidgeon, N.J. Framing and communicating climate change: The effects of distance and outcome frame manipulations. *Glob. Environ. Chang.* **2010**, *20*, 656–667. [CrossRef]
17. Vollstedt, B.; Koerth, J.; Tsakiris, M.; Nieskens, N.; Vafeidis, A.T. Co-production of climate services: A story map for future coastal flooding for the city of Flensburg. *Clim. Serv.* **2021**, *22*, 100225. [CrossRef]
18. Hagemeier-Klose, M.; Wagner, K.J.N.H.; Sciences, E.S. Evaluation of flood hazard maps in print and web mapping services as information tools in flood risk communication. *Nat. Hazards Earth Syst. Sci.* **2009**, *9*, 563–574. [CrossRef]
19. Fuchs, S.; Spachinger, K.; Dorner, W.; Rochman, J.; Serrhini, K.J.E.H. Evaluating cartographic design in flood risk mapping. *Environ. Hazards* **2009**, *8*, 52–70. [CrossRef]
20. Oubennaceur, K.; Chokmani, K.; Gauthier, Y.; Ratte-Fortin, C.; Homayouni, S.; Toussaint, J.-P. Flood Risk Assessment under Climate Change: The Petite Nation River Watershed. *Climate* **2021**, *9*, 125. [CrossRef]
21. Chokmani, K.; Oubennaceur, K.; Tanguy, M.; Poulin, J.; Gauthier, Y.; Latapie, R.; Bernier, M. The Use of Remotely Sensed Information within a Flood Risk Management and Analysis Tool (GARI). In Proceedings of the IGARSS 2019-2019 IEEE International Geoscience and Remote Sensing Symposium, Yokohama, Japan, 28 July–2 August 2019; pp. 4636–4639.
22. Chokmani, K.; Tanguy, M.; Oubennaceur, K.; Poulin, J.; Gauthier, Y.; Bernier, M. GARI: A tool for flood risk management and analysis, beyond mapping of flooded areas. In *EGU General Assembly Conference Abstracts*; EGU: Vienna, Austria, 2018; p. 9672.
23. Zheng, X.; Tarboton, D.G.; Maidment, D.R.; Liu, Y.Y.; Passalacqua, P. River channel geometry and rating curve estimation using height above the nearest drainage. *JAWRA J. Am. Water Resour. Assoc.* **2018**, *54*, 785–806. [CrossRef]
24. Centre D'expertise Hydrique du Québec. *Atlas Hydroclimatique du Québec Méridional: Impact des Changements Climatiques sur les Régimes de Crue, D'étiage et D'hydraulicité à L'horizon 2050*; Gouvernement du Québec: Québec, QC, Canada, 2013.
25. Walshe, N. Using ArcGIS online story maps. *Teach. Geogr.* **2016**, *41*, 115–117.
26. Kerski, J. *Types of Story Maps. Spatial Thinking in Environmental Contexts: Maps, Archives and Timelines*; CRC Press: Boca Raton, FL, USA, 2019.

MDPI
St. Alban-Anlage 66
4052 Basel
Switzerland
Tel. +41 61 683 77 34
Fax +41 61 302 89 18
www.mdpi.com

Hydrology Editorial Office
E-mail: hydrology@mdpi.com
www.mdpi.com/journal/hydrology

www.ingramcontent.com/pod-product-compliance
Lightning Source LLC
LaVergne TN
LVHW070627170726
843515LV00004B/203

* 9 7 8 3 0 3 6 5 3 7 7 8 8 *